T0319367

POWER ELECTRONIC CONVERTERS FOR MICROGRIDS

POWER ELECTRONIC CONVERTERS FOR MICROGRIDS

Suleiman M. Sharkh

University of Southampton, United Kingdom

Mohammad A. Abusara

University of Exeter, United Kingdom

Georgios I. Orfanoudakis

University of Southampton, United Kingdom

Babar Hussain

Pakistan Institute of Engineering and Applied Sciences, Pakistan

This edition first published 2014
© 2014 John Wiley & Sons Singapore Pte. Ltd.

Registered office
John Wiley & Sons Singapore Pte. Ltd., 1 Fusionopolis Walk, #07-01 Solaris South Tower, Singapore 138628.

For details of our global editorial offices, for customer services and for information about how to apply for permission to reuse the copyright material in this book please see our website at www.wiley.com.

All rights reserved. No part of this publication may be reproduced, stored in a retrieval system or transmitted, in any form or by any means, electronic, mechanical, photocopying, recording, scanning, or otherwise, except as expressly permitted by law, without either the prior written permission of the Publisher, or authorization through payment of the appropriate photocopy fee to the Copyright Clearance Center. Requests for permission should be addressed to the Publisher, John Wiley & Sons Singapore Pte. Ltd., 1 Fusionopolis Walk, #07-01 Solaris South Tower, Singapore 138628, tel: 65-66438000, fax: 65-66438008, email: enquiry@wiley.com.

Wiley also publishes its books in a variety of electronic formats. Some content that appears in print may not be available in electronic books.

Designations used by companies to distinguish their products are often claimed as trademarks. All brand names and product names used in this book are trade names, service marks, trademarks or registered trademarks of their respective owners. The Publisher is not associated with any product or vendor mentioned in this book. This publication is designed to provide accurate and authoritative information in regard to the subject matter covered. It is sold on the understanding that the Publisher is not engaged in rendering professional services. If professional advice or other expert assistance is required, the services of a competent professional should be sought.

MATLAB® is a trademark of The MathWorks, Inc. and is used with permission. The MathWorks does not warrant the accuracy of the text or exercises in this book. This book's use or discussion of MATLAB® software or related products does not constitute endorsement or sponsorship by The MathWorks of a particular pedagogical approach or particular use of the MATLAB® software.

Limit of Liability/Disclaimer of Warranty: While the publisher and author have used their best efforts in preparing this book, they make no representations or warranties with respect to the accuracy or completeness of the contents of this book and specifically disclaim any implied warranties of merchantability or fitness for a particular purpose. It is sold on the understanding that the publisher is not engaged in rendering professional services and neither the publisher nor the author shall be liable for damages arising herefrom. If professional advice or other expert assistance is required, the services of a competent professional should be sought.

Library of Congress Cataloging-in-Publication Data has been applied for.

A catalogue record for this book is available from the British Library.

ISBN 978-0-470-82403-0

Set in 11/13pt Times by Laserwords Private Limited, Chennai, India
Printed and bound in Singapore by Markono Print Media Pte Ltd

1   2014

Contents

About the Authors

Suleiman M. Sharkh obtained his BEng and PhD degrees in Electrical Engineering from the University of Southampton in 1990 and 1994, respectively. He is currently the Head of the Electro-Mechanical Research Group at the University of Southampton. He is also the Managing Director of HiT Systems Ltd, and a visiting Professor at the Beijing Institute of Technology and Beijing Jiaotong University.

He has 20 years research experience in the field of electrical and electromagnetic systems, including electric switches, power electronics, electrical machines, control systems, and characterization and management of advanced batteries. To date he has published about 150 publications. He has obtained research grant income of about £2M from the Research Councils and industry since 1998. He has supervised 11 PhD students to completion and is currently supervising 5 PhD students. He is an established doctoral external examiner in the UK and abroad, including Europe, China, and Australia. His research has contributed to the development of a number of commercial products, including rim driven marine thrusters (TSL Technology Ltd), down-hole submersible motors for drilling and pumping oil wells (TSL Technology Ltd), sensorless brushless DC motor controllers (TSL Technology), power electronic converters for microgrids (Bowman Power Systems and TSL Technology), high-speed PM alternators for Rankine cycle and gas microturbine energy recovery systems (TSL Technology, Bowman Power Systems, and Freepower), and battery management systems (Reap Systems Ltd).

He was the winner of The Engineer Energy Innovation and Technology Award that was presented at the Royal Society London in October 2008 for his work on novel rim driven marine thrusters and turbine generators, which are produced commercially under licence by TSL Technology Ltd. He was also awarded the Faraday SPARKS award in 2002. He is a past committee member of the UK Magnetics Society, a member of the IET and a Chartered Engineer.

Mohammad A. Abusara received the BEng degree from Birzeit University, Palestine, in 2000, and the PhD degree from the University of Southampton, UK, in 2004, both in electrical engineering. From 2003 to 2010, he was with Bowman Power Group, Southampton, UK, responsible for research and development of digital control of power electronics for distributed energy resources, hybrid vehicles, and machines and drives. He is currently a Senior Lecturer in Renewable Energy at the University of Exeter, UK.

Georgios I. Orfanoudakis received his MEng in Electrical Engineering and Computer Science from the National Technical University of Athens (NTUA), Greece, in 2007, and his MSc in Sustainable Energy Technologies from the University of Southampton, UK, in 2008. He then joined the Electro-Mechanical Research Group at the University of Southampton and obtained his PhD in 2013. His research focused on the modulation and DC-link capacitor sizing of three-level inverters. Since October 2012 he is working as a Research Associate in a Knowledge Transfer Partnership (KTP) with the University of Southampton and TSL technology Ltd., performing R&D work on inverters for motor drive applications. Dr Orfanoudakis is a member of the IEEE Power Electronics Society.

Babar Hussain received the BSc degree in electrical engineering from the University of Engineering and Technology, Taxila, Pakistan, in 1995 and the PhD degree in electrical engineering from the University of Southampton, Southampton, UK, in 2011. He has more than 10 years experience in the electric power sector. His major research interests include protection of distribution networks with distributed generation, power quality, and control of grid-connected inverters.

Preface

Microgrids and distributed generation (DG), including renewable sources and energy storage, can help overcome power system capacity limitations, improve efficiency, reduce emissions, and manage the variability of renewable sources. A key component of such a system is the power electronic interface between a generator or an energy storage system, and the grid. Such an interface needs to be capable of performing several functions, including injection of high quality current into the grid to meet national standards; charging and discharging energy storage systems in a controlled manner; anti-islanding protection to disconnect from the grid when the mains are lost; and continuing to supply critical loads when the grid is lost.

The aim of this book is to provide an in-depth coverage of specific topics related to power electronic converters for microgrids, focusing on three-phase converters in the range 50–250 kW. It also discusses the important problem of protection of distribution networks, including microgrids and DG. The book is intended as a textbook for graduating students with an electrical engineering background who wish to work or do research in this field.

Chapter 1 presents a review of the state of the art and future challenges of power electronic converters used in microgrids. Chapter 2 describes the structure and modulation strategies of the conventional two-level and three-level neutral point clamped (NPC) and cascaded H-Bridge (CHB) converter topologies. Chapter 3 discusses the sizing of DC-link capacitors in two-level and three-level inverters, based on expressions for the rms values and the harmonic spectrum of the capacitor current. Chapter 4 investigates semiconductor and DC-link capacitor losses in two-level and three-level inverters, and presents a comparison between the different topologies. Chapter 5 investigates the problem of low-frequency voltage oscillations that appear at the neutral point of an NPC converter, and proposes an algorithm for minimizing these oscillations. Chapter 6 discusses the design and practical implementation of a digital current controller for a three-phase two-level voltage source grid-connected inverter with an LCL output filter. Chapter 7 discusses the design and control of a three-phase voltage source grid-connected interleaved inverter and describes its practical implementation. Chapter 8 discusses the design and practical implementation of a repetitive controller for an interleaved grid-connected inverter. Chapter 9 discusses the design and practical implementation of a line interactive UPS (uninterruptible power supply) system capable of seamlessly transferring between grid-connected to stand-alone modes in parallel with other sources, as well as managing the charging and discharging of the battery. Chapter 10 discusses protection issues and challenges arising from the integration of microgrids and DG into the grid. Chapter 11 discusses the problem of recloser–fuse coordination in a distribution network including microgrids and DG, and proposes an adaptive fuse saving scheme that takes into account the status of the DG. There are also two appendices. Appendix A gives some background material on SVM (space vector modulation) for NPC converters and describes the MATLAB®/Simulink models and programs used to carry out the simulations in Chapter 5. Appendix B includes the MATLAB® code used to numerically calculate the DC-link capacitor rms current and voltage ripple.

Acknowledgments

The authors would like to thank and acknowledge the valuable support of TSL Technology Ltd and Bowman Power Group Ltd who funded some of the research included in this book. In particular they wish to thank Dr Mike Yuratich at TSL and Mr John Lyons at Bowman for their support and help over the last 15 years. They also wish to acknowledge the contributions of Dr Zahrul Faizi Hussain and Dr Mohsin Jamil to some of the material presented in this book. Thanks are also due to the UK Engineering and Physical Science Research Council (EPSRC) for supporting Dr Orfanoudakis's PhD. We wish also to express special appreciation to the staff at Wiley, especially James Murphy, Clarissa Lim, and Shelley Chow for their support, patience, and trust in this project.

1

Introduction

Fossil fuels are running out and current centralized power generation plants using these fuels are inefficient with a significant amount of energy lost as heat to the environment. These plants also produce harmful emissions and greenhouse gases. Furthermore, existing power systems, especially in developing countries, suffer from several limitations, such as the high cost of expansion and efficiency improvement limits within the existing grid infrastructure. Renewable energy sources can help address these issues, but their variable nature poses challenges to their integration within the grid.

Distributed generators (DGs), including renewable sources, within microgrids can help overcome power system capacity limitations, improve efficiency, reduce emissions, and manage the variability of renewable sources. A microgrid, a relatively new concept, is a zone within the main grid where a cluster of electrical loads and small microgeneration systems, such as solar cells, fuel cells, wind turbine, and small combined heat and power (CHP) systems, exist together under an embedded management and control system, with the option of energy storage. Other benefits of generating power close to electrical loads include the use of waste heat locally, saving the cost of upgrading the grid to supply more power from central plants, reducing transmission losses, and creating opportunities for increasing competition in the sector, which can stimulate innovation and reduce consumer prices [1, 2].

Power electronic converters are used in microgrids to control the flow of power and convert it into suitable DC or AC form. Different types of converter are needed to perform the many functions within a microgrid, but it is not the aim of this chapter or this book to review all of these possible types of converter, many of which are covered in textbooks and other publications [3]. The book will primarily focus on converters used to connect DG systems, including microCHP and renewable energy sources, to an AC grid or to local AC loads, as illustrated in Figure 1.1. They convert DC (from photovoltaic cells [4], batteries, fuel cells [5]) or variable frequency AC (wind and marine turbine [6]) into 50/60 Hz AC power that is injected into the grid and/or used

Power Electronic Converters for Microgrids, First Edition. Suleiman M. Sharkh, Mohammad A. Abusara, Georgios I. Orfanoudakis and Babar Hussain.
© 2014 John Wiley & Sons, Ltd. Published 2014 by John Wiley & Sons, Ltd.
Companion Website: www.wiley.com/go/sharkh

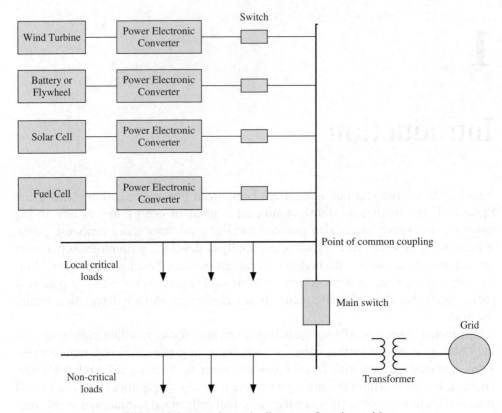

Figure 1.1   A schematic diagram of a microgrid

to supply local loads. Converters are also used to connect flywheel energy storage systems or high-speed microturbine generators to the grid.

1.1   Modes of Operation of Microgrid Converters

Normally, converters are used to connect DG systems in parallel with the grid or other sources, but it may be useful for the converters to continue functioning in stand-alone mode when the other sources become unavailable, in order to supply critical loads. Converters connected to batteries or other storage devices will also need to be bidirectional to charge and discharge these devices.

1.1.1   Grid Connection Mode

In this mode of operation, the converter connects the power source in parallel with other sources to supply local loads and possibly feed power into the main grid. Parallel connection of embedded generators is governed by national standards [7–9]. The standards require that the embedded generator should not regulate or oppose the voltage at the point of common coupling, and that the current fed into the grid should be of high quality with upper limits on its total harmonic distortion (THD). There is also a limit on the maximum DC current injected into the grid.

The power injected into the grid can be controlled either by direct control of the current fed into the grid [10] or by controlling the power angle [11]. In the latter case, the voltage is controlled to be sinusoidal. However, using power angle control without directly controlling the output current may not be effective at reducing the output current THD when the grid voltage is highly distorted. But this is also an issue in the case of electric machine generators which effectively use power angle control. This raises the question of whether it is reasonable to specify current THD limits, regardless of the quality of the utility voltage.

In practice, the converter output current or voltage needs to be synchronized with the grid, which is usually achieved by using a phase-locked loop or grid voltage zero crossing detection [12]. The standards also require that embedded generators, including power electronic converters, should incorporate an anti-islanding feature, so that they are disconnected from the point of common coupling when the grid power is lost. There are many anti-islanding techniques, the most common being the rate of change of frequency (RoCoF) technique [13].

1.1.2   Stand-Alone Mode

It may be desirable for the converter to continue to supply a critical local load when the main grid is disconnected, for example, by the anti-islanding protection system. In this stand-alone mode the converter needs to maintain constant voltage and frequency, regardless of load imbalance or the quality of the current, which can be highly distorted if the load is nonlinear.

A situation may arise in a microgrid, disconnected from the main grid, where two or more power electronic converters switch to stand-alone mode to supply a critical load. In this case, these converters need to share the load equitably. Equitable sharing of load by parallel connected converters operating in stand-alone mode requires additional control. There are several methods for parallel connection, which can be broadly classified into two categories: (i) frequency and voltage droop method [14] and (ii) master-slave method, whereby one of the converters acts as a master, setting the frequency and voltage, and communicating to the other converters their share of the load [15].

1.1.3   Battery Charging Mode

In a microgrid, storage batteries, or other energy storage devices are needed to handle disturbances and fast load changes [16]. In other words, energy storage is needed to accommodate the variations of available power generation and demand, thus improving the reliability of the microgrid. The power electronic converter could then be used as a battery charger.

1.2   Converter Topologies

Most of the current commercially available power electronic converters used for grid connection are based on the voltage-source two-level PWM (pulse width modulation) inverter, as illustrated in Figure 1.2 [10, 17]. An LCL filter is commonly used, but L filters have been also used [18, 19]. An LCL filter is smaller in size compared to a simple L filter, but it requires a more complex control system to manage the LC resonance. Additionally, the impedance of L_2C in Figure 1.2 tends to be relatively low, which provides an easy path for current harmonics to flow from the grid. This can cause the THD of the current to increase beyond permitted limits in cases where the grid voltage THD is relatively high. Ideally, this drawback could be overcome by increasing the feedback controller gain in a current controlled grid connected converter, but this can prove to be difficult to achieve in practice while maintaining good stability [20].

Other filter topologies have also been proposed. For example, Guoqiao *et al.* [21], proposed an LCCL filter arrangement, feeding back the current measured between the two capacitors. By selecting the values of the capacitors to match the inductor values, the closed loop transfer function of the system becomes non-resonant, which helps to improve the performance of the controller, as discussed in Section 1.4.

The size and cost of the filter can be very significant. Filter size can be reduced by either increasing the switching frequency of the converter or reducing the converter voltage step changes. However, the switching frequency, which is limited by losses in the power electronic devices, tends to reduce as the power ratings of the devices and the converters increase. This means that high power two-level converters could have disproportionately large filters.

Alterative converter topologies, which can help reduce the size of the filter, have been the subject of recent research. Multi-level converters have been proposed, including the neutral point clamped (NPC) inverter shown in Figure 1.3a [22], and the

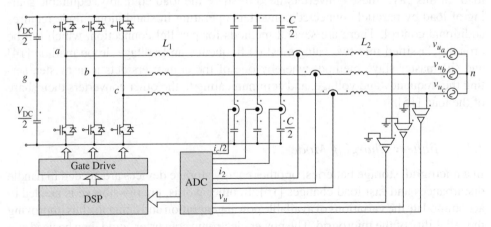

Figure 1.2   Twolevel grid-connected inverter with LCL filter

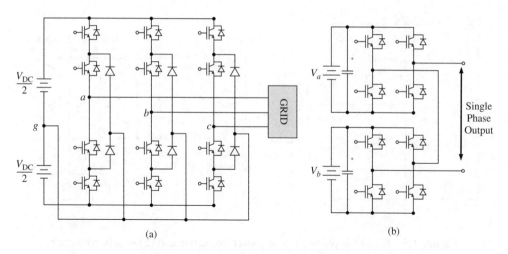

(a)   (b)

Figure 1.3   Multi-level voltage source inverter (a) NPC and (b) cascaded

cascaded converter shown in Figure 1.3b (only one phase is shown) [23]. Multi-level converters have the advantage of reducing the voltage step changes, and hence the size and the cost of the main filter inductor for given current ripple, at the expense of increased complexity and cost of the power electronics and control components [24].

An alternative to the multi-level converter is to use an interleaved converter topology, as illustrated in Figure 1.4, which shows a converter that has two channels. A grid-connected converter based on this topology, with six channels, has already

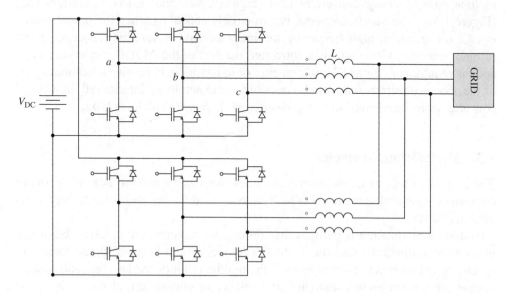

Figure 1.4   Interleaved converter with two channels

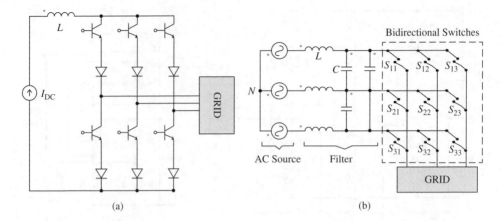

Figure 1.5   (a) Three-phase current source converter and (b) matrix converter

been designed, built, and tested by the authors (see Chapter 7). Interleaving is a form
of paralleling technique where the switching instants are phase shifted over a switching period. By introducing an equal phase shift between parallel power stages, the
output filter capacitor ripple is reduced due to the ripple cancellation effect [25, 26].
Additionally, by using smaller low current devices, it is possible to switch at high
frequency, and therefore the inductors and overall filter size would be smaller than
a single channel inverter. The number of channels in an interleaved converter is a
compromise between complexity and filter size.

Other possible converter topologies that have been considered for this application
include current source converters [27] (Figure 1.5a), and matrix converters [28]
(Figure 1.5b). The matrix converter is particularly appealing when the power source
is AC, for example, high frequency turbine generator or variable frequency wind
turbine generator. Using a matrix converter, the cost of the AC/DC conversion stage
and the requirement for a DC link capacitor or inductor could be saved. Combinations
of the above converters may also be possible, for example, an interleaved multi-level
converter or an interleaved matrix converter, perhaps with soft switching.

1.3   Modulation Strategies

There are a variety of modulation techniques that can be used in power electronic
converters in general, including PWM, hysteresis modulation, and pulse density modulation (PDM).

Hysteresis modulation is perhaps the simplest to implement in practice, but it has
many shortcomings: (i) variable switching frequency and hence a spread harmonic
spectrum, (ii) increased current error if the middle point of the DC-link and system
neutral are not connected, and (iii) poor quality of output current means it has to
be used with other techniques, such as repetitive feedback, to improve the output

current quality [17]. Due to these shortcomings, it is not preferred for microgrid inverters where high quality output current and good transient response are essential requirements.

Not surprisingly, most grid connected converters use PWM, either carrier-based or using space vector modulation (SVM), which has the advantage of ease of implementation using a microprocessor. Third harmonic injection is often used with sinusoidal PWM to reduce the required DC-link voltage headroom. SVM strategies have also been developed to minimize switching losses or eliminate certain harmonics. In multi-level and interleaved converters, the number of SVM states increases significantly, with many redundant states, which creates further opportunities for device switching strategies to reduce or redistribute the losses within the converter and eliminate certain harmonics [29, 30].

PDM is another possible modulation technique. It is not commonly used in conventional converters but it has been used in high-frequency (150 kHz) converters used for induction heating [31]. The potential for this modulation strategy is yet to be explored in the context of converters for grid connection applications.

1.4   Control and System Issues

The control system of a grid-connected converter needs to cater for the different possible operating modes that were discussed earlier in Section 1.1. In the grid-connected mode, either a maximum power tracking system or the user will specify the power and power factor to be injected into the grid. The control system needs then to translate that into a reference demand current, if the output current into the grid is to be controlled. Alternatively, the controller needs to determine the output converter voltage and power angle, if power flow into the grid is based on power angle control. The reference signals need to be synchronized with the grid, as mentioned earlier.

It is common to use the d-q transformation to translate the measured AC signals of voltage and current to DC. This in effect is equivalent to a resonant controller, thus providing steady state error at the fundamental frequency, which simplifies controller design and implementation using a microprocessor-based controller [32]. However, such an approach assumes that the measured signals are pure sinusoids and that the grid is balanced, which in practice is often not the case. A slight imbalance as well as harmonic distortions are often present, which act as disturbances that cause a deterioration of the output current THD.

The alternative is to have a separate controller for each phase, with direct control of the sinusoidal output current, as discussed in Chapters 5 and 6. But direct feedback of the output grid current of an LCL filter on its own can be inherently unstable (unless the filter has some resistive damping), and it is necessary to have another feedback loop of the capacitor current (see Figures 1.2 and 1.6) or the current in the main inductor L_1 [10, 17, 33]. One of the challenging aspects of this controller structure is that it is not possible to have a high outer loop gain using a simple compensation or

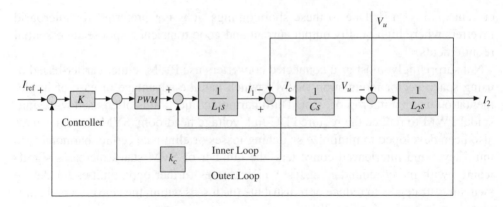

Figure 1.6 Two feedback loop structure

PID (proportional-integral-derivative) controller. Resonant controllers [34] and virtual inductance [35] (see also Chapter 9) could be used to help increase the outer loop gain and, hence, improve disturbance rejection. Other types of controller, including optimal control strategies [36], state-feedback approach [20], and sliding mode controllers [37], have also been proposed.

An alternative is to use feedforward schemes to compensate for grid voltage harmonics [17] or inverter dead time [38], at the expense of extra complexity and cost. Repetitive or cyclic feedback has also been proposed: the output current is compared with the demanded current on a cycle by cycle basis and, accordingly, the effective reference current demanded from the inverter is modified to compensate for the disturbances. However, repetitive feedback was found in practice to lack robustness and to be sensitive to parameter uncertainty [39]. Chapter 8 discusses the design of a robust repetitive feedback control system for a grid-connected inverter.

Guoqiao et al. [19] proposed splitting the capacitor into two capacitors in parallel, and adding a minor feedback loop of the current measured after the first capacitor, as mentioned earlier. By carefully selecting the capacitor values, the transfer function of the system can be reduced to be first order, which helps to mitigate the filter resonance problem and enables the outer loop gain to be increased, thus improving disturbance rejection. In practice, it may be difficult to find capacitors with the right values and the uncertainty in the filter values will make the condition for resonance elimination difficult to achieve.

For simplicity and guaranteed stability, many authors [40–42] choose to control the inverter current before the filter rather than the grid current. Such systems can, however, suffer from problems resulting from filter resonance, such as underdamped transient response oscillations, large overshoot, and oscillations induced by utility harmonics near the resonance frequency, in addition to poor utility voltage harmonic disturbance rejection. Blasko and Kaura [43] proposed a lead-lag compensating loop in the filter capacitor voltage to actively dampen filter resonance. Resistors connected in series with capacitors or inductors are also sometimes used, although this is usually not an efficient option in a classical two-level inverter. However, as discussed in Chapter 7, passive damping using resistors in series with the filter capacitors can be an efficient, low cost method in an interleaved inverter.

Computational time delay, when a digital controller is used can be significant and could affect system stability. To maintain system stability, a time delay compensation scheme may need to be implemented [15], as discussed in Chapter 6, but with faster DSPs (digital signal processors), FPGAs (field programmable gate arrays), and microcontrollers, this is becoming less of an issue. Care also needs to be taken when generating the reference sinusoidal signal using a look-up table in a microprocessor, as using an insufficient number of samples may cause sub-harmonic distortion [44].

The controller also needs to incorporate an anti-islanding protection feature, and needs to seamlessly switch from grid-connected mode to stand-alone mode to supply critical loads, in parallel with other converters. This could mean switching from a current control strategy to a voltage control strategy, which can be challenging to implement. An alternative is to adopt a voltage and power angle (i.e., frequency) control strategy for all modes of operation, which will make the transition between different modes seamless, as discussed in Chapter 9. As mentioned earlier, however, this control method does not directly control the current injected into the grid and meeting the THD standards may be challenging if the grid voltage THD is relatively high. Chapter 9 proposes using a virtual inductance approach to help improve the THD.

Often, metering of power is incorporated as a function in commercial controllers, which can be remotely interrogated via Ethernet, CAN (controller area network) bus, GPRS, or wireless connection.

1.5   Future Challenges and Solutions

The cost and size of the converter and filter components remain an issue. This, in principle, may be addressed by increasing the frequency and using either multilevel or interleaved topologies, as discussed earlier. However, further research is needed to establish the optimum trade-off between power electronic devices and filter components.

It may also be beneficial to take an overall system approach when designing the converter, rather than considering the design of the converter in isolation. For example, a cascaded multilevel converter requires multiple isolated DC-links, which requires a multi-tap transformer or complex schemes using flying capacitors. One alternative could be to design the electric machine (of say a wind generator) to have multiple sets of three-phase coils, isolated from each other, and connected to independent rectifiers to generate the isolated DC-links for the multilevel converter. Another could be to design the isolated three-phase windings to have a phase shift, for example, a 30° phase shift between two sets of three-phase windings, which can be connected to a

12-pulse rectifier, which reduces the size and cost of the DC-link capacitor. Similarly, in a photovoltaic system, the solar cells may be connected to form isolated DC sources.

The proliferation of inverter interfaced DG units is already raising issues related to coordination of protection relays, both in grid-connected and stand-alone modes [45]. Chapter 10 discusses these issues and proposes some solutions.

The standards governing these converters are still evolving and practical implementation is continuously giving rise to new issues that need to be thought about and regulated.

References

1. Sharkh, S.M., Arnold, R.J., Kohler, J. *et al.* (2006) Can microgrids make a major contribution to UK energy supply? *Renewable and Sustainable Energy Reviews*, **10**, 78–127.
2. Barnes, M., Kondoh, J., Asano, H. *et al.* (2007) Real-world microgrids – an overview. IEEE International Conference on System Systems Engineering, SoSE '07, pp. 1–8.
3. Biczel, P. (2007) Power electronic converters in DC microgrid. Conference on Compatibility in Power Electronics, CPE '07, pp. 1–6.
4. Rong-Jong, W. and Wen-Hung, W. (2007) Design of grid-connected photovoltaic generation system with high step-up converter and sliding-mode inverter control. IEEE International Conference on Control Applications, CCA 2007, pp. 1179–1184.
5. Gaiceanu, M. (2007) Inverter control for three-phase grid connected fuel cell power system. Conference on Compatibility in Power Electronics, CPE '07, pp. 1–6.
6. Zhiling, Q. and Guozhu, C. (2007) Study and design of grid connected inverter for 2 MW wind turbine. IEEE 42nd IAS Annual Meeting, Industry Applications Conference, pp. 165–170.
7. Engineering Recommendation G59/1 (1991) Recommendations for the Connection of Embedded Generating Plant to the Regional Electricity Companies Distribution Systems. Electricity Association (Engineering Services).
8. IEEE (2000) Std 929-2000. IEEE Recommended Practice for Utility Interface of Photovoltaic (PV) Systems, IEEE.
9. Basso, T.S. and DeBlasio, R. (2004) IEEE 1547 series of standards: interconnection issues. *IEEE Transactions on Power Electronics*, **19**, 1159–1162.
10. Sharkh, S.M. and Abu-Sara, M. (2004) Digital current control of utility connected two-level and three-level PWM voltage source inverters. *European Power Electronic Journal*, **14**, 13–18.
11. Mohan, N., Undeland, T.M. and Robbins, W.P. (2006) *Power Electronics Converters, Applications, and Design*, 3rd edn, John Wiley and Sons, Inc., Hoboken, NJ.
12. Arruda, L.N., Silva, S.M., and B.J.C. Filho (2001) PLL structures for utility connected systems. EEE 36th IAS Annual Meeting, Industry Applications Conference, Vol. 4, pp. 2655–2660.
13. Mahat, P., Zhe, C., and Bak-Jensen, B. (2008) Review of islanding detection methods for distributed generation. Third International Conference on Electric Utility Deregulation and Restructuring and Power Technologies, DRPT 2008, pp. 2743–2748.
14. De Brabandere, K., Bolsens, B., Van den Keybus, J. *et al.* (2007) A voltage and frequency droop control method for parallel inverters. *IEEE Transactions on Power Electronics*, **22**, 1107–1115.

15. Abu-Sara, M. (2004) Digital control of utility and parallel connected three-phase PWM inverters. PhD thesis. University of Southampton.

16. Lopes, J.A.P., Moreira, C.L. and Madureira, A.G. (2006) Defining control strategies for MicroGrids islanded operation. *IEEE Transactions on Power Systems*, **21**, 916–924.

17. Sharkh, S.M., Abu-Sara, M., and Hussein, Z.F. (2001) Current control of utility -connected DC-AC three-phase voltage-source inverters using repetitive feedback. European Power Electronic Conference, EPE 2001, Graz, Austria.

18. Eric, W. and Lehn, P.W. (2006) Digital current control of a voltage source converter with active damping of LCL resonance. *IEEE Transactions on Power Electronics*, **21**, 1364–1373.

19. Guoqiao, S., Dehong, X., Luping, C. and Xuancai, Z. (2008) An improved control strategy for grid-connected voltage source inverters with an LCL filter. *IEEE Transactions on Power Electronics*, **23**, 1899–1906.

20. Gabe, I.J., Massing, J.R., Montagner, V.F., and Pinheiro, H. (2007) Stability analysis of grid-connected voltage source inverters with LCL-filters using partial state feedback. European Conference on Power Electronics and Applications, pp. 1–10.

21. Guoqiao, S., Dehong, X., Danji, X., and Xiaoming,Y. (2006) An improved control strategy for grid-connected voltage source inverters with a LCL filter. IEEE 21st Annual Applied Power Electronics Conference and Exposition, APEC '06, p. 7.

22. Bouhali, O., Francois, B., Saudemont, C., and Berkouk, E.M. (2006) Practical power control design of a NPC multilevel inverter for grid connection of a renewable energy plant based on a FESS and a Wind generator. 32nd Annual Conference on IEEE Industrial Electronics, IECON 2006, pp. 4291–4296.

23. Selvaraj, J. and Rahim, N.A. (2009) Multilevel inverter for grid-connected PV system employing digital PI controller. *IEEE Transactions on Industrial Electronics*, **56**, 149–158.

24. Ikonen, M., Laakkonen, O., and Kettunen, M. (2005) Two-Level and Three-Level Converter Comparison in Wind Power Application, www.elkraft.ntnu.no/smola2005/Topics /15.pdf, Finland (accessed 23 October 2013).

25. Xunbo, F., Chunliang, E., Jianlin, L., and Honghua, X. (2008) Modeling and simulation of parallel-operation grid-connected inverter. IEEE International Conference on Industrial Technology, pp. 1–6.

26. Zhang, M.T., Jovanovic, M.M. and Lee, F.C.Y. (1998) Analysis and evaluation of interleaving techniques in forward converters. *IEEE Transactions on Power Electronics*, **13**, 690–698.

27. Lee, S.-H., Song, S.-G., Park, S.-J. *et al.* (2008) Grid-connected photovoltaic system using current-source inverter. *Solar Energy*, **82**, 411–419.

28. Barakati, S.M., Kazerani, M. and Chen, X. (2005) A new wind turbine generation system based on matrix converter. *IEEE Power Engineering Society General Meeting*, **3**, 2083–2089.

29. Kazmierkowski, M.P. and Malesani, L. (1998) Current control techniques for three-phase voltage-source PWM converters a survey. *IEEE Transactions on Industrial Electronics*, **45**, 691–703.

30. Kojabadi, H.M., Bin, Y., Gadoura, I.A. *et al.* (2006) A novel DSP-based current-controlled PWM strategy for single phase grid connected inverters. *IEEE Transactions on Power Electronics*, **21**, 985–993.

31. Nguyen-Quang, N., Stone, D.A., Bingham, C., and Foster, M.P. (2007) Comparison of single-phase matrix converter and H-bridge converter for radio frequency induction heating. European Conference on Power Electronics and Applications, pp. 1–9.
32. Twining, E. and Holmes, D.G. (2003) Grid current regulation of a three-phase voltage source inverter with an LCL input filter. *IEEE Transactions on Power Electronics*, **18**, 888–895.
33. Twining, E. and Holmes, D.G. (2002) Grid current regulation of a three-phase voltage source inverter with an LCL input filter. IEEE 33rd Annual Power Electronics Specialists Conference, pp. 1189–1194.
34. Timbus, A.V., Ciobotaru, M., Teodorescu, R., and Blaabjerg, F. (2006) Adaptive resonant controller for grid-connected converters in distributed power generation systems. Annual IEEE Applied Power Electronics Conference and Exposition, p. 6.
35. Jou, H.L., Chiang, W.J., and Wu, J.C. Virtual inductor-based islanding detection method for grid-connected power inverter of distributed power generation system. IET Renewable Power Generation Conference, Vol. 1, pp. 175-181, 2007.
36. Lee, K.-J., Park, N.-J., and Hyun, D.-S. (2007) Optimal current controller in a three-phase grid connected inverter with an LCL filter. Hong Kong Conference on Control.
37. Kim, I.-S. (2006) Sliding mode controller for the single-phase grid-connected photovoltaic system. *Applied Energy*, **83**, 1101–1115.
38. Qingrong, Z. and Liuchen, C. (2005) Improved current controller based on SVPWM for three-phase grid-connected voltage source inverters. IEEE 36th Power Electronics Specialists Conference, pp. 2912–2917.
39. Weiss, G., Qing-Chang, Z., Green, T.C. and Jun, L. (2004) H∞ repetitive control of DC-AC converters in microgrids. *IEEE Transactions on Power Electronics*, **19**, 219–230.
40. Lindgren, M. and Svensson, J. (1998) Control of a voltage-source converter connected to the grid through an LCL-filter-application to active filtering. IEEE Power Electronics Specialists Conference, pp. 229–235.
41. Liserre, M., Blaabjerg, F. and Hansen, S. (2005) Design and control of an LCL-filter-based three-phase active rectifier. *IEEE Transactions on Industry Applications*, **41**, 1281–1291.
42. Prodanovic, M. and Green, T.C. (2003) Control and filter design of three-phase inverters for high power quality grid connection. *IEEE Transactions on Power Electronics*, **18**, 373–380.
43. Blasko, V. and Kaura, V. (1997) A novel control to actively damp resonance in input LC filter of a three-phase voltage source converter. *IEEE Transactions on Industry Applications*, **33**, 542–550.
44. Abeyasekera, T., Johnson, C.M., Atkinson, D.J. and Armstrong, M. (2003) Elimination of subharmonics in direct look-up table (DLT) sine wave reference generators for low-cost microprocessor-controlled inverters. *IEEE Transactions on Power Electronics*, **18**, 1315–1321.
45. Feero, W.E., Dawson, D.C., and Stevens, J. (2002) White paper on Protection Issues of The MicroGrid Concept. Consortium for Electric Reliability Technology Solutions, March 2002.

2

Converter Topologies

This chapter covers the structure, modulation, and modeling in MATLAB®-Simulink of the conventional two-level (2L), and the three-level (3L) neutral point clamped (NPC) and cascaded H-bridge (CHB) converter topologies.

2.1 Topologies

This section presents the structure of 2L, NPC, and CHB converters, together with their switching states and conduction paths.

2.1.1 The Two-Level Converter

The three-phase 2L converter, illustrated in Figure 2.1, consists of three legs (a, b, c), each comprising two switching modules ($V_{1a} - V_{2a}$, $V_{1b} - V_{2b}$, $V_{1c} - V_{2c}$). The modules are formed by an active switch (e.g. IGBT (insulated gate bipolar transistor), IGCT (insulated gate commutated thyristor), MOSFET (metal oxide semiconductor field-effect transistor), etc.) and a diode, connected anti-parallel to the switch. The three converter legs are connected across a common DC-link capacitor (C), which provides a low-inductance path for the rapidly varying currents through the modules.

The active switching of the modules is controlled (i.e., turned on/off) by gating signals given to the module drivers. The gating signals for modules V_{1x} and V_{2x} will be symbolized as g_{1x} and g_{2x}, respectively, where x can be a, b, or c. Each gating signal can be equal either to 0 (switch is off) or to 1 (switch is on). However, g_{1x} and g_{2x} should never be made equal to 1 at the same time because this would short-circuit the converter's DC-link capacitor. During the converter operation, g_{1x} and g_{2x} for each leg are, therefore, complementary (apart from short intervals of dead-time, during which both signals are set to 0).

Figure 2.2 illustrates the possible switching states (s_x) of each converter leg, defined by the allowed combinations of the gating signals. In the 2L converter, each leg has

Power Electronic Converters for Microgrids, First Edition. Suleiman M. Sharkh, Mohammad A. Abusara, Georgios I. Orfanoudakis and Babar Hussain.
© 2014 John Wiley & Sons, Ltd. Published 2014 by John Wiley & Sons, Ltd.
Companion Website: www.wiley.com/go/sharkh

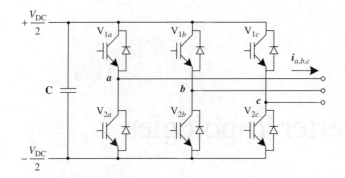

Figure 2.1 Two-level converter topology

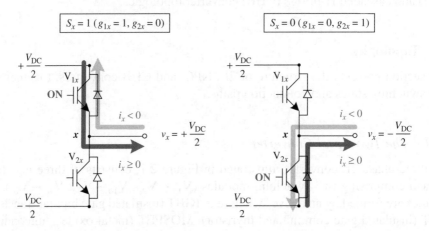

Figure 2.2 Switching states and conduction paths for a leg of the two-level converter

two possible switching states, $s_x = 1$ or 0, outputting a phase voltage of $+V_{DC}/2$ and $-V_{DC}/2$, respectively. For a given state, the conduction path changes according to the direction of the current. It can be noticed, though, that the phase voltage is solely dependent on the gating signals; it is not affected by the phase current.

2.1.2 The NPC Converter

Each NPC converter comprises four switching modules ($V_{1x} - V_{4x}$) and two diodes (D_{5x} and D_{6x}), as shown in Figure 2.3. The converter's DC-link consists of two capacitors (C_1 and C_2), connected at the converter's neutral point (NP). The topology takes its name from the fact that the (clamping) diodes D_{5x} and D_{6x} clamp the voltage of the points found between the outer and inner switching modules to the NP voltage. This results in each converter module switching across the voltage of one of the DC-link

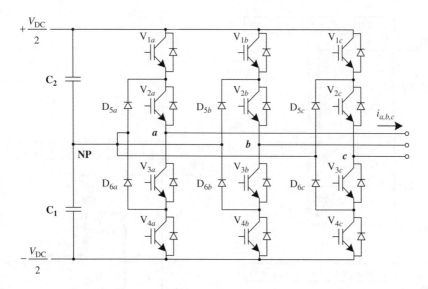

Figure 2.3   NPC converter topology

capacitors. If the capacitors are balanced, the switching voltage of each module is therefore $V_{DC}/2$.

The gating signals are provided to each leg of the NPC converter to turn on two adjacent modules at any time. Turning on the upper/lower three adjacent modules of a leg would short-circuit the upper/lower DC-link capacitor, respectively, while turning on all four modules, would short-circuit the whole DC-link. Consequently, each leg can only be found at three different switching states, $s_x = 2, 1$, or 0, illustrated in Figure 2.4. The phase voltage is equal to $+V_{DC}/2$, v_{NP}, and $-V_{DC}/2$, respectively. Again, the conduction path changes according to the direction of the current but the phase voltage remains unaffected.

2.1.3   The CHB Converter

Unlike the 2L and 3L NPC, the CHB converter is not comprised of three legs connected to a common DC-link. Instead, it is based on three H-bridge cells (single-phase, 3L converters), each having its own, isolated DC-link. As compared to the 2L and NPC, each cell of the CHB needs to have half the DC-link voltage for the converter to be able to generate the same (fundamental) output voltage. This has been illustrated in Figure 2.5, by setting the cells' DC-link voltages to $V_{DC}/2$ in place of V_{DC}. As for the case of the NPC converter, the module switching voltage is therefore $V_{DC}/2$. The three cells are connected at a common neutral, n.

Each cell can have four different switching states, resulting from turning on two of the cell modules that do not belong to the same leg. Figure 2.6 illustrates the allowed

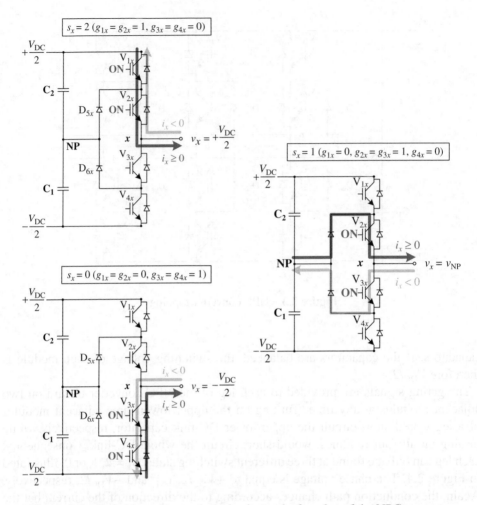

Figure 2.4 Switching states and conduction paths for a leg of the NPC converter

switching states, together with the respective conduction paths. As for the NPC converter, states 2 and 0 produce a phase voltage of $+V_{DC}/2$ and $-V_{DC}/2$, respectively. Both of the remaining states, 1a and 1b, produce a phase voltage of zero, since they connect the cell output (x) to the neutral (n).

2.2 Pulse Width Modulation Strategies

Pulse width modulation (PWM) generates pulsed voltage waveforms whose average (over a switching cycle) is equal to the desired reference signals. Depending on the way they are implemented, PWM strategies can be categorized as carrier-based or space-vector modulation (SVM) strategies.

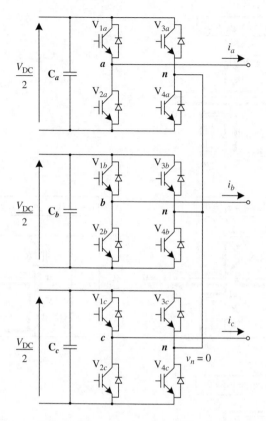

Figure 2.5   CHB converter topology

2.2.1   *Carrier-Based Strategies*

Carrier-based strategies utilize a set of reference and carrier waveforms to generate the converter PWM voltages. Three reference waveforms, one for each converter leg, are used in three-phase converters. A reference waveform provides the desired value of output voltage for the respective phase, normalized with respect to $V_{DC}/2$. For the case of sinusoidal pulse width modulation (SPWM), the reference waveforms for phases a, b, and c, are respectively given by the following equations:

$$v_{a,\text{ref}} = M \cos(\theta) \tag{2.1}$$

$$v_{b,\text{ref}} = M \cos\left(\theta - \frac{2\pi}{3}\right) \tag{2.2}$$

$$v_{c,\text{ref}} = M \cos\left(\theta + \frac{2\pi}{3}\right) \tag{2.3}$$

where M is the converter modulation index and θ is the reference angle.

Figure 2.7 illustrates the carrier, reference, and PWM voltage waveforms for phase a, for a 2L converter modulated using SPWM and unity modulation index. The reference

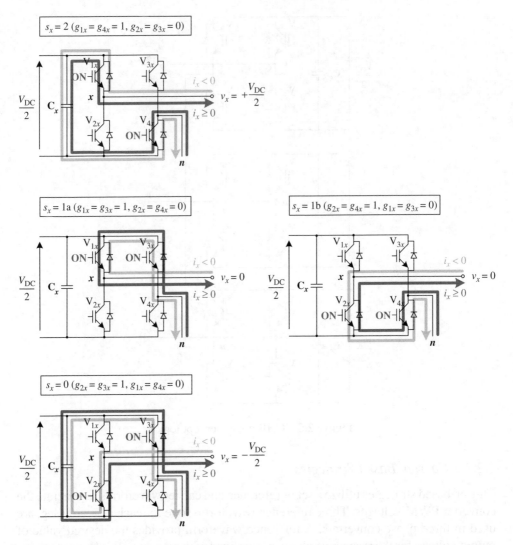

Figure 2.6 Switching states and conduction paths for a leg (H-bridge) of the CHB converter

waveforms in a 2L converter are compared with a single, common carrier waveform to determine the width of the generated pulses. The carrier waveform is commonly triangular. Its peak values are +1 and −1, and its frequency is equal to the switching frequency of the converter.

While the value of a reference is higher than that of the carrier waveform, the state, s_x, of the respective phase is set to 1, and the phase voltage becomes equal to $+V_{DC}/2$; otherwise, s_x is set to 0 and the phase voltage becomes $-V_{DC}/2$. The resulting PWM waveform for phase x has a duty cycle of

$$\delta_x = \frac{1}{2}(1 + v_{x,\text{ref}}) \qquad (2.4)$$

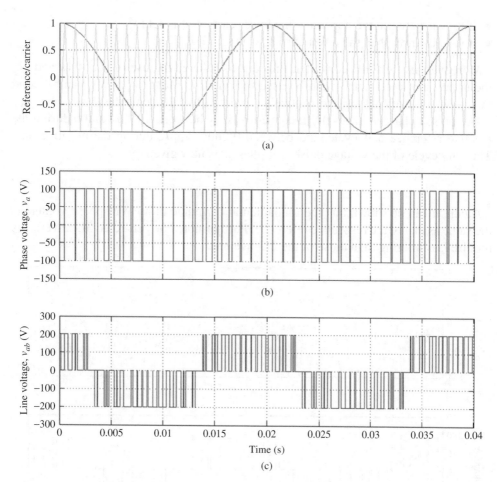

Figure 2.7 (a) Reference and carrier waveforms for phase a, (b) phase voltage v_a, and (c) line voltage v_{ab} for a two-level converter, assuming $V_{DC} = 200$ V, $f = 50$ Hz, and $f_s = 1$ kHz

It can be shown that over a period of the carrier, T_s, the area of this waveform, normalized with respect to (w.r.t.) $V_{DC}/2$, is the same as the area of $v_{x,ref}$. This ensures that the fundamental harmonic component of the PWM waveform is the same as that of the reference waveform. Filtering of the higher-order harmonics turns the PWM into the desired sinusoidal voltage waveform.

The carrier-based SPWM strategy for 3L converters uses the three reference waveforms described by Equations 2.1–2.3, and two carrier waveforms, arranged as shown in Figure 2.8. The carrier waveforms are in phase for the so-called phase disposition (PD) PWM strategies, improving the PWM voltage harmonic spectra [1]. The state of each leg or cell of a 3L NPC or CHB converter, respectively, is determined as follows:

• If $v_{x,ref}$ is greater than the upper carrier, s_x is set to 2.

- If $v_{x,\text{ref}}$ is between the upper and the lower carrier, s_x is set to 1.
- If $v_{x,\text{ref}}$ is lower than the lower carrier, s_x is set to 0.

Moreover, in the CHB converter, in order to balance the use of states 1a and 1b, the upper and lower carrier can be used to switch the first $(V_{1x} - V_{2x})$ and second $(V_{3x} - V_{4x})$ leg of each cell, respectively (similarly to a 2L converter).

The pulsed phase voltages in 3L converters vary between $+V_{\text{DC}}/2$ and 0 during the positive reference half cycle, and between 0 and $-V_{\text{DC}}/2$ during the negative one. The duty cycle of the voltage pulses for phase x is now given by

$$\delta_x = v_{x,\text{ref}} \tag{2.5}$$

Again, it can be shown that the fundamental harmonic of a PWM phase voltage is the same as that of the respective reference. Furthermore, the generated 3L waveforms require less filtering than those of the 2L inverter.

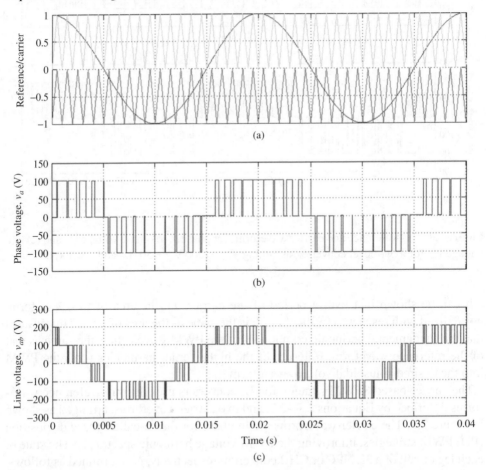

Figure 2.8 (a) Reference and carrier waveforms for phase a, (b) phase voltage v_a, and (c) line voltage v_{ab} for a three-level converter, assuming $V_{\text{DC}} = 200\,\text{V}, f = 50\,\text{Hz}$, and $f_s = 1\,\text{kHz}$

For both 2L and 3L converters, the sinusoidal phase references of the SPWM strategy result in sinusoidal line voltages, which are supplied to the converter load. Sinusoidal line voltages, however, can also be generated by non-sinusoidal phase references if the latter are modified by a common-mode component, *cm*:

$$v_{a,\text{ref}} = M\cos(\theta) + cm \tag{2.6}$$

$$v_{b,\text{ref}} = M\cos\left(\theta - \frac{2\pi}{3}\right) + cm \tag{2.7}$$

$$v_{c,\text{ref}} = M\cos\left(\theta + \frac{2\pi}{3}\right) + cm \tag{2.8}$$

The line voltages are accordingly given by:

$$v_{ab,\text{ref}} = v_{a,\text{ref}} - v_{b,\text{ref}} = M\left[\cos(\theta) - \cos\left(\theta - \frac{2\pi}{3}\right)\right] = \sqrt{3}M\cos\left(\theta + \frac{\pi}{6}\right) \tag{2.9}$$

$$v_{bc,\text{ref}} = v_{b,\text{ref}} - v_{c,\text{ref}} = M\left[\cos\left(\theta - \frac{2\pi}{3}\right) - \cos\left(\theta + \frac{2\pi}{3}\right)\right] = \sqrt{3}M\cos\left(\theta - \frac{2\pi}{3} + \frac{\pi}{6}\right) \tag{2.10}$$

$$v_{ca,\text{ref}} = v_{c,\text{ref}} - v_{a,\text{ref}} = M\left[\cos\left(\theta + \frac{2\pi}{3}\right) - \cos(\theta)\right] = \sqrt{3}M\cos\left(\theta + \frac{2\pi}{3} + \frac{\pi}{6}\right) \tag{2.11}$$

It can be seen that the line voltages are not affected by the insertion of a common-mode voltage component. The amplitude of the line voltages, however, (which is equal to $\sqrt{3}M$) can be increased by using a common-mode voltage that allows an increase in the maximum value of M. Namely, the maximum value of M for the case of SPWM is 1, since higher values lead to over-modulation and introduce low-frequency voltage distortion. An appropriate common-mode signal, on the other hand, can be added to the reference voltages to keep them in the range of ± 1 while M increases beyond 1. It can be shown that the maximum value that M can take in this way is

$$M_{\max} = \frac{2}{\sqrt{3}} \approx 1.1547 \tag{2.12}$$

which leads to a respective maximum line voltage of $2V_{\text{DC}}$ peak-peak.

A typical common-mode waveform used for the above purpose is described by

$$cm_{3h} = -\frac{1}{6}\cos(3\theta) \tag{2.13}$$

corresponding to a method known as "third harmonic injection" (SPWM + third harmonic). The waveform of $v_{a,\text{ref}}$ and the PWM voltages are shown in Figure 2.9 for this strategy, at unity modulation index. It can be seen from Figure 2.9a that the modulation index can now increase further, without leading to over-modulation of the converter.

2.2.2 SVM Strategies

SVM differs from carrier-based implementations of modulation strategies, since the former (i) is based on numerical calculations instead of waveform intersections and (ii) works directly with line voltages. Starting with the 2L converter, the two states available for each leg lead to the functional diagram shown in Figure 2.10 for the entire three-phase converter.

The converter can therefore have $2^3 = 8$ switching states. Each converter state is represented by a space vector on the complex plane, given by the following transformation:

$$\mathbf{V} = \frac{2}{\sqrt{3}}\left(s_a + s_b e^{j\frac{2\pi}{3}} + s_c e^{-j\frac{2\pi}{3}}\right) \tag{2.14}$$

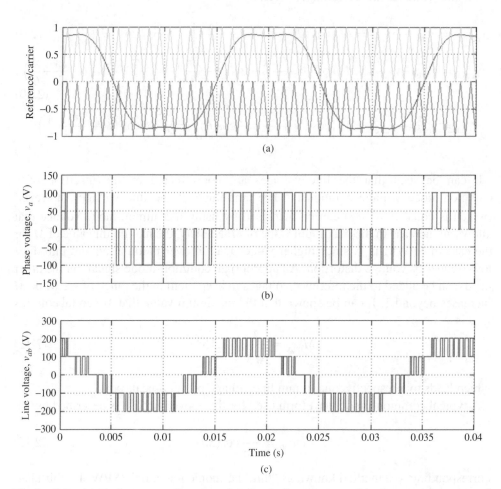

(a)

(b)

(c)

Figure 2.9 (a) Reference and carrier waveforms for phase a, (b) phase voltage v_a, and (c) line voltage v_{ab} for a three-level converter modulated by SPWM + third harmonic, assuming $V_{DC} = 200$ V, $f = 50$ Hz, and $f_s = 1$ kHz

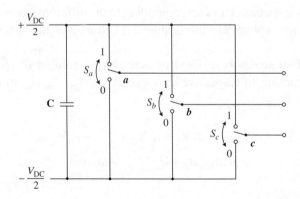

Figure 2.10 Functional diagram of the two-level converter

Table 2.1 Space vectors and line voltages for the two-level converter states

State ($s_a s_b s_c$)	Space vector	Line voltages (v_{ab}, v_{bc}, v_{ca})
000	0	0, 0, 0
100	$2/\sqrt{3} \cdot e^{j0}$	V_{DC}, 0, $-V_{DC}$
110	$2/\sqrt{3} \cdot e^{j\pi/3}$	0, V_{DC}, $-V_{DC}$
010	$2/\sqrt{3} \cdot e^{j2\pi/3}$	$-V_{DC}$, V_{DC}, 0
011	$2/\sqrt{3} \cdot e^{j\pi}$	$-V_{DC}$, 0, V_{DC}
001	$2/\sqrt{3} \cdot e^{j4\pi/3}$	0, $-V_{DC}$, V_{DC}
101	$2/\sqrt{3} \cdot e^{j5\pi/3}$	V_{DC}, $-V_{DC}$, 0
111	0	0, 0, 0

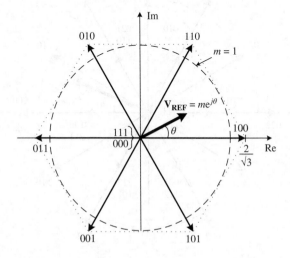

Figure 2.11 Space vector diagram for the two-level converter

Table 2.1 lists the space vectors corresponding to the different states and relates them to the converter line voltages, while Figure 2.11 illustrates the space vectors on the complex plane.

PWM in SVM strategies is realized by activating a number of space vectors, V_1, V_2, ..., V_n, according to respective duty cycles d_1, d_2, ..., d_n, to create a voltage reference vector, V_{REF}:

$$V_{REF} = d_1 V_1 + d_2 V_2 + ... + d_n V_n$$

$$\text{with} \quad d_1 + d_2 + ... + d_n = 1 \tag{2.15}$$

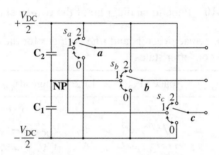

Figure 2.12   Functional diagram of the NPC converter

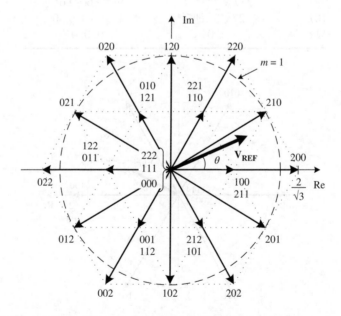

Figure 2.13   Space vector diagram for three-level converters

The reference vector represents the three converter line voltages on the SV plane and is defined as follows:

$$\mathbf{V}_{REF} = \frac{1}{\sqrt{3}} \left(v_{a,ref} + v_{b,ref}e^{j\frac{2\pi}{3}} + v_{c,ref}e^{-j\frac{2\pi}{3}} \right) \tag{2.16}$$

Assuming that the voltage references are given by Equations 2.6–2.8, $\mathbf{V}_{REF}$ can be shown to be equal to

$$\mathbf{V}_{REF} = \frac{\sqrt{3}}{2}M \cdot e^{j\theta} = m \cdot e^{j\theta} \tag{2.17}$$

where m will be the symbol for the modulation index for SVM strategies. As shown in Figure 2.11, m is equal to 1 at the limit of the linear (not over-) modulation region, and relates to M as determined by

$$m = \frac{\sqrt{3}}{2}M \tag{2.18}$$

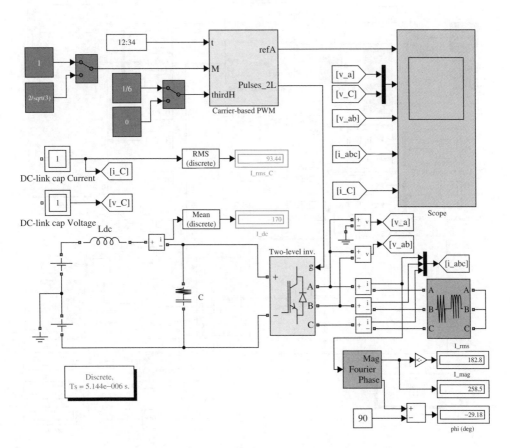

Figure 2.14   Simulink model for the two-level converter

SVM strategies can operate with $m=1$, or equivalently $M=M_{max}$ – see Equation 2.12 – generating the maximum possible amplitude of line voltage.

Figure 2.12 illustrates the functional diagram of a 3L NPC converter. This converter can be found at $3^3 = 27$ switching states. The respective space vectors are now derived by

$$V = \frac{1}{\sqrt{3}} \left(s_a + s_b e^{j\frac{2\pi}{3}} + s_c e^{-j\frac{2\pi}{3}} \right) \tag{2.19}$$

and are shown in Figure 2.13. The SV diagram is the same for the 3L CHB converter. It can be noticed that there are pairs of small vectors (e.g., 100-211) and a triplet of zero vectors (000-111-222) that share the same position on the SV plane. The same is true for the two zero vectors at the middle of the SV diagram for the 2L converter (Figure 2.11). This property of the SV diagrams is essential for creating different SV modulation strategies, since vectors with the same position on the SV plane can be used alternatively to create the reference vector according to Equation 2.15. The vector

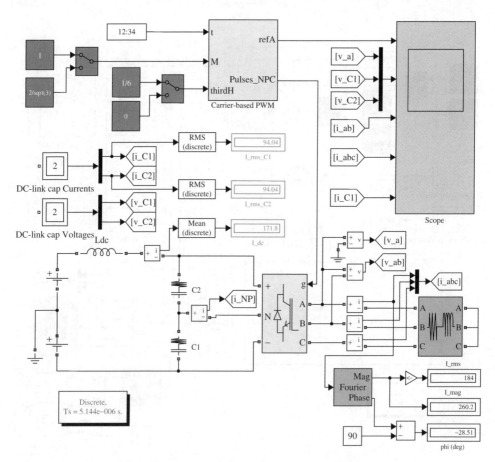

Figure 2.15   Simulink model for the NPC converter

selection, on the other hand, determines the common-mode voltage of the converter, correlating SVM to carrier-based strategies, as described in [2–5].

2.3 Modeling

MATLAB®-Simulink SimPowerSystems toolbox is a powerful power electronic simulation tool. It is, however, easier to use Embedded MATLAB® functions to implement (code) modulation strategies. Figures 2.14–2.16 illustrate top level models of the 2L, and 3L NPC and CHB converters. In the models are shown:

- The converters and their (carrier-based) modulation functions, which provide the gating signals to the converter modules.
- Manual controls for adjusting the converter modulation index and switching between SPWM and SPWM + third harmonic.
- The converter load and measurements associated with it.

Additional Simulink models and MATLAB® code are included in Appendix A.

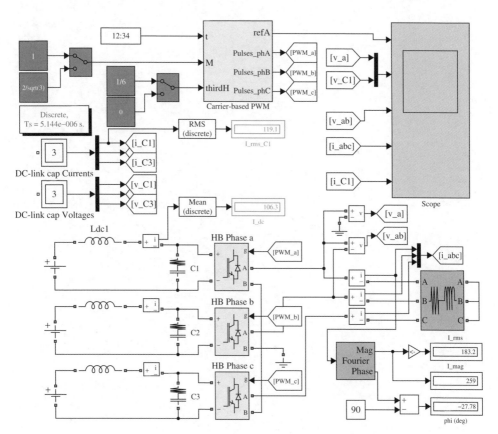

Figure 2.16   Simulink model for the CHB converter

References

1. Holmes, D.G. and Lipo, T.A. (2003) *Pulse Width Modulation for Power Converters*, IEEE Press, Piscataway, NJ.
2. Wang, C. and Li, Y. (2010) Analysis and calculation of zero-sequence voltage considering neutral-point potential balancing in three-level NPC converters. *IEEE Transactions on Industrial Electronics*, **57** (7), 2262–2271.
3. Da-peng, C., Wen-xiang, S., Hui, X.I. *et al.* (2009) Research on zero-sequence signal of space vector modulation for three-level neutral-point-clamped inverter based on vector diagram partition. IEEE 6th International Power Electronics and Motion Control Conference.
4. Nho, N.V. and Youn, M.-J. (2006) Comprehensive study on space-vector-PWM and carrier-based-PWM correlation in multilevel invertors. *IEEE Proceedings on Electric Power Applications*, **153** (1), 149–158.
5. Pou, J., Zaragoza, J., Ceballos, S. *et al.* (2012) A carrier-based PWM strategy with zero-sequence voltage injection for a three-level neutral-point-clamped converter. *IEEE Transactions on Power Electronics*, **27** (2), 642–651.

3

DC-Link Capacitor Current and Sizing in NPC and CHB Inverters

3.1 Introduction

During the last decades, the presence of multilevel (ML) inverters has been steadily increasing in a variety of applications in the manufacturing, transport, energy, mining, and other industries [1, 2]. An essential part of ML inverter design is the selection of DC-link capacitors. The capacitors are a sensitive element of the inverter and a common source of failures. Capacitor lifetime is highly affected by thermal as well as overvoltage stresses, both of which can be estimated based on an analysis of the capacitor current [3, 4].

Thermal stress occurs due to losses on the DC-link capacitor's equivalent series resistance (ESR). The rms value of the current flowing through a DC-link capacitor can provide a first approximation for its losses. The literature contains rms expressions for the capacitor current of the two-level [5–7] and (single-phase) three-level cascaded H-bridge (CHB) inverters [8]. Use of these expressions for loss estimation assumes a fixed ESR value. However, the ESR is a function of the frequency of the capacitor current [3, 4]. Since the current of a DC-link capacitor comprises several harmonics located at different frequencies, it is necessary to determine the rms values of the capacitor current harmonics and use the appropriate value of ESR for each harmonic. For the two-level inverter, DC-link current harmonics have been derived in [9], while a general analytical method for calculating DC-link current harmonics in inverters has been proposed in [10].

Overvoltage stress may occur due to low-frequency (LF) capacitor voltage oscillations (ripple). In the two-level inverter, such oscillations do not appear under balanced load conditions. This is not the case, however, in three-level inverters. A plot of the amplitude of the capacitor voltage ripple in the three-level neutral point clamped (NPC) inverter [11] can be found in [12], for the case of sinusoidal pulse width modulation (SPWM). Moreover, capacitor-balancing techniques for the

Power Electronic Converters for Microgrids, First Edition. Suleiman M. Sharkh, Mohammad A. Abusara, Georgios I. Orfanoudakis and Babar Hussain.
© 2014 John Wiley & Sons, Ltd. Published 2014 by John Wiley & Sons, Ltd.
Companion Website: www.wiley.com/go/sharkh

NPC inverter, which achieve smaller ripple amplitudes, are described in [13–16]. Regarding the CHB inverter topology, estimates for the capacitor voltage ripple can be found in [17], again for the case of SPWM.

This chapter presents a systematic analysis of the DC-link capacitor current in three-level NPC and CHB inverters, which provides the basis for DC-link capacitor sizing in these topologies. Methods for analyzing the two-level inverter DC-link capacitor current are extended to three-level inverters, to estimate the capacitor rms current and derive its harmonic spectrum. A new numerical approach for calculating the rms value and LF harmonics of the capacitor current is also proposed. Unlike existing methods, the proposed approach has the advantage of being easily adaptable to different modulation strategies and higher-level inverters. Results based on this approach are presented for a number of common modulation strategies.

The chapter is structured as follows: Section 3.2 gives a description of the parameters considered during DC-link capacitor sizing, explaining their relation to the capacitor current. Section 3.3 derives expressions for the rms value of the DC-link capacitor current in three-level NPC and CHB topologies [18], while Section 3.4 presents a harmonic analysis of this current [19]. Section 3.5 describes the proposed numerical method for deriving the current rms value and the amplitude of voltage ripple for the DC-link capacitors of the two topologies. Then, Section 3.6 validates the results of the previous sections using simulations in MATLAB®-Simulink. Finally, Section 3.7 compares the examined three-level topologies with respect to their capacitor requirements, discusses the accuracy and applicability of the presented methods, and describes the application of the proposed approach to higher-level inverters.

3.2 Inverter DC-Link Capacitor Sizing

The selection of inverter DC-link capacitors is determined by the required voltage, (ripple) current, and capacitance ratings, as explained below:

- The capacitor voltage rating is typically higher than the operating DC-link voltage, to account for voltage oscillations and other effects such as input (grid) voltage fluctuations or transitory regenerative operation of the inverter.
- The ripple (rms) current rating, $I_{C,max}$, is a way of expressing the affordable limit for capacitor losses, $P_{C,max}$. $I_{C,max}$ is commonly given in capacitor datasheets [3, 4] for a certain value of ESR, R_C, as

$$P_{C,max} = R_C \cdot I_{C,max}^2 \qquad (3.1)$$

According to the above, a calculation of the DC-link capacitor rms current in a certain inverter application can give a first estimate for the required ripple current rating. However, for electrolytic capacitors, the value of the ESR varies with the frequency of the capacitor current. A typical ESR–frequency characteristic is illustrated in Figure 3.1 [4].

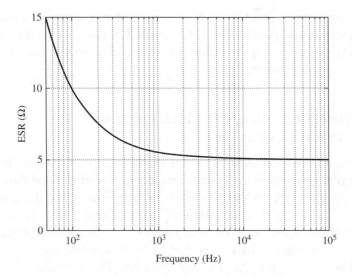

Figure 3.1   ESR–frequency characteristic of a 4.7 mF/450 V capacitor. (Reproduced from [4].)

Hence, if more than one current harmonic h, with amplitudes I_h (rms values $I_{h,\text{rms}}$) and frequencies f_h flow through the capacitor, the losses should be calculated using

$$P_C = \sum_h R_C(f_h) \cdot I_{h,\text{rms}}^2 \tag{3.2}$$

where $R_C(f_h)$ stands for the value of ESR at frequency f_h. Then, the capacitor's ripple current rating can be selected so that P_C does not exceed $P_{C,\text{max}}$.

- The required capacitance is determined by the affordable amplitude of the DC-link voltage oscillations ($\Delta V_{C,\text{max}}/2$), which is also dependent on the capacitor current harmonics. Assuming that, in the worst case, all these harmonics (or equivalently all peaks of voltage harmonics) are in phase, the required capacitance is given by

$$C \geq \frac{1}{\Delta V_{C,\text{max}}/2} \cdot \sum_h \frac{I_h}{2\pi f_h} \tag{3.3}$$

According to this equation, high-frequency (HF) capacitor current harmonics, which appear due to the switching (PWM, pulse width modulation) operation of the inverter, have a small effect on the capacitor voltage ripple. LF harmonics, on the other hand, located at multiples of the inverter fundamental frequency, can increase the required capacitance, since they give rise to LF capacitor voltage oscillations.

It is noted that this study focuses on the effect of the inverter on the rating of the DC-link capacitor. Increased capacitance can, therefore, be required for decreasing the DC-link capacitor voltage ripple caused by the inverter front end (e.g., a rectifier), or for other purposes, such as for voltage support during a temporary loss of input power. Inversely, less capacitance may be required in cases where the power source (e.g., a battery) is capable of providing part of the LF capacitor current as this will decrease the capacitor voltage ripple.

3.3 Analytical Derivation of DC-Link Capacitor Current RMS Expressions

In this section the method used in [5] for the derivation of the two-level inverter capacitor current rms expression is summarized and extended to the three-level NPC and CHB topologies. The method considers each inverter IGBT-diode module as a switch that, while on, carries the current of the respective phase. If the sum of the currents through the upper switches of an inverter is i_d, then the DC component of i_d is (assumed to be) supplied by the inverter DC source, while the AC component is filtered, and hence carried by the DC-link capacitor. The rms value of the capacitor current, $I_{C,rms}$, is calculated using the average (DC) and rms values of i_d, $I_{d,DC}$, and $I_{d,rms}$, respectively as

$$I_{C,rms} = \sqrt{I_{d,rms}^2 - I_{d,DC}^2} \tag{3.4}$$

The calculation of the average and rms values for current i_d is based on the analysis of its transitions within a single switching period. If i_d is equal to $i_{d,int1}$, $i_{d,int2}$, $\ldots$, $i_{d,intk}$, during k time intervals T_{int1}, T_{int2}, $\ldots$, T_{intk}, respectively, then its average and rms values during a switching period, T_s, centered at reference angle θ, are given by

$$i_{d,DC}(\theta) = \frac{1}{T_s}\left(\sum_k T_{intk} \cdot i_{d,intk}\right) = \sum_k \delta_{intk} \cdot i_{d,intk} \tag{3.5}$$

$$i_{d,rms}^2(\theta) = \frac{1}{T_s}\left(\sum_k T_{intk} \cdot i_{d,intk}^2\right) = \sum_k \delta_{intk} \cdot i_{d,intk}^2 \tag{3.6}$$

(where the interval duty cycles, δ_{intk}, and currents, $i_{d,intk}$, are also functions of θ, but $\delta_{intk}(\theta)$ and $i_{d,intk}(\theta)$ are used in place of them, respectively, for reasons of conciseness). The interval duty cycles δ_{intk} and respective currents $i_{d,intk}$ are also functions of θ. The average and rms values of i_d over a fundamental period are obtained using the following expressions:

$$I_{d,DC} = \frac{1}{2\pi}\int_0^{2p} i_{d,DC}(\theta)d\theta \tag{3.7}$$

$$I_{d,rms} = \sqrt{\frac{1}{2\pi}\int_0^{2\pi} i_{d,rms}^2(\theta)\, d\theta} \tag{3.8}$$

If the expressions for $i_{d,DC}(\theta)$ and $i_{d,rms}(\theta)$ change during parts (sectors) of the fundamental cycle, the above expressions are written as sums of integrals for the different sectors.

3.3.1 NPC Inverter

For the three-level NPC inverter, the current i_d is shown in Figure 3.2a. The AC component of i_d is the current of the upper capacitor of the DC-link (C_2). Assuming SPWM, the (normalized w.r.t. $V_{DC}/2$) phase reference voltages, $v_{x,ref}$, shown in Figure 3.3, and the respective duty cycles, δ_x, are given by Equations 3.9 and 3.10 below,

$$v_{x,ref} = M\sin(\theta + \theta_x) \tag{3.9}$$

$$\delta_x = v_{x,ref} = M\sin(\theta + \theta_x) \tag{3.10}$$

where M is the inverter modulation index, while θ_x for phases a, b, and c is equal to θ_a, θ_b, and θ_c, respectively:

$$\theta_a = 0, \theta_b = \frac{2\pi}{3}, \theta_c = -\frac{2\pi}{3} \tag{3.11}$$

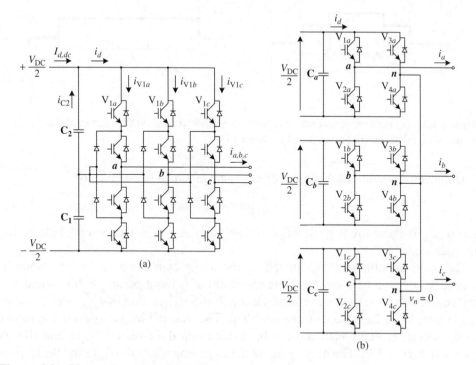

Figure 3.2 Three-phase three-level (a) neutral-point-clamped and (b) cascaded H-bridge inverter

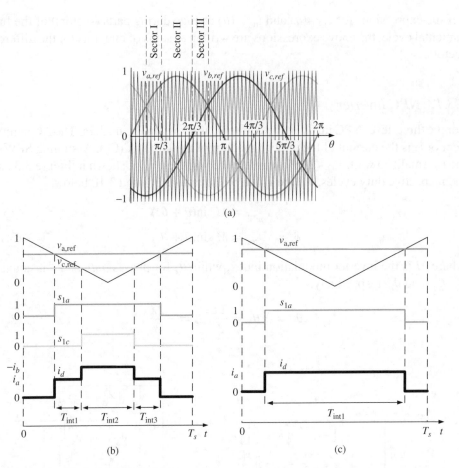

Figure 3.3 (a) Reference and upper carrier waveforms for the NPC inverter, and analysis of a switching cycle for (b) Sector I and (c) Sector II

Moreover, the sinusoidal phase currents i_x are assumed to be given by

$$i_x = I_{pk} \sin(\theta + \theta_x - \phi) \qquad (3.12)$$

where I_{pk} is the current peak value (magnitude) and ϕ is the power angle of the inverter load.

The upper modules are modulated by the upper carrier waveform, as shown in Figure 3.3a. It can be shown that the waveform of i_d has a period of $T/3$, where T is the period of a reference waveform. Hence, for the derivation of $I_{d,DC}$ and $I_{d,rms}$, an angle interval of $2\pi/3$ needs to be analyzed. The selected interval covers the values of θ between $\pi/6$ and $5\pi/6$ and can be divided into three sectors (I, II, and III), as shown in Figure 3.3a. The duty cycles of the switching intervals (δ_{intk}) and the respective i_d currents ($i_{d,intk}$) change between Sectors I and II, as shown in Figure 3.3b,c. In the figure, s_{1x} represents the state of module V_{1x} (0 for OFF and 1 for ON).

Tables 3.1 and 3.2 present the expressions of the above duty cycles in a concise form that can be used according to Equations 3.7 and 3.8 to derive the DC and rms values of i_d for each sector. The resulting expressions are shown in Table 3.3 (Sector III can be analyzed similarly to Sector I).

Since the expressions for $i_{d,DC}(\theta)$ and $i_{d,rms}(\theta)$ change during sectors of the fundamental cycle, Expressions 3.7 and 3.8 take the form of Equations 3.13 and 3.14, respectively,

$$I_{d,DC\text{-}NPC} = \frac{3}{2\pi} \left(\int_{\pi/6}^{\pi/3} i_{d,DC,I\text{-}NPC}(\theta)\,d\theta + \int_{\pi/3}^{2\pi/3} i_{d,DC,II\text{-}NPC}(\theta)d\theta \right.$$
$$\left. + \int_{2\pi/3}^{5\pi/6} i_{d,DC,III\text{-}NPC}(\theta)\,d\theta \right) \tag{3.13}$$

$$I_{d,rms\text{-}NPC} = \sqrt{\frac{3}{2\pi}\left[\begin{pmatrix} \int_{\pi/6}^{\pi/3} i^2_{d,rms,I\text{-}NPC}(\theta)\,d\theta + \int_{\pi/3}^{2\pi/3} i^2_{d,rms,II\text{-}NPC}(\theta)d\theta \\ + \int_{2\pi/3}^{5\pi/6} i^2_{d,rms,III\text{-}NPC}(\theta)d\theta \end{pmatrix} \right]} \tag{3.14}$$

Table 3.1 Switching intervals for Sector I of the NPC inverter

Total duration	Total duty cycle	Current i_d
$T_{int1} + T_{int3}$	$\delta_a - \delta_c$	i_a
T_{int2}	δ_c	$-i_b$

Table 3.2 Switching intervals for Sector II of the NPC inverter

Total duration	Total duty cycle	Current i_d
T_{int1}	δ_a	i_a

Table 3.3 Expressions for $i_{d,DC}(\theta)$ and $i_{d,rms}(\theta)$ for the three sectors of the NPC inverter

Sector I	Sector II	Sector III
$i_{d,DC,I\text{-}NPC}(\theta) =$ $(\delta_a - \delta_c)i_a + \delta_c(-i_b)$ $i^2_{d,rms,I\text{-}NPC}(\theta) =$ $(\delta_a - \delta_c)i_a^2 + \delta_c(-i_b)^2$	$i_{d,DC,II\text{-}NPC}(\theta) = \delta_a i_a$ $i^2_{d,rms,II\text{-}NPC}(\theta) = \delta_a i_a^2$	$i_{d,DC,III\text{-}NPC}(\theta) =$ $(\delta_a - \delta_b)i_a + \delta_b(-i_c)$ $i^2_{d,rms,III\text{-}NPC}(\theta) =$ $(\delta_a - \delta_b)i_a^2 + \delta_b(-i_c)^2$

which results in:

$$I_{d,DC-NPC} = \frac{3}{4} M I_{pk} \cos \phi \tag{3.15}$$

$$I_{d,rms-NPC} = I_{pk} \sqrt{\frac{\sqrt{3}M}{4\pi} \left(\cos^2\phi + \frac{1}{4} \right)} \tag{3.16}$$

Finally, according to Equation 3.4, the rms current expression for the upper capacitor (C_2) of the three-level NPC inverter is shown below:

$$I_{C,rms-NPC} = I_{pk} \sqrt{\frac{M}{2} \left[\frac{\sqrt{3}}{2\pi} + \left(\frac{2\sqrt{3}}{\pi} - \frac{9}{8} M \right) \cos^2(\phi) \right]} \tag{3.17}$$

Due to symmetry (between the positive and negative half cycles) of the reference and carrier waveforms, the expression for the lower DC-link capacitor (C_1) is identical.

3.3.2   CHB Inverter

The derivation of the DC-link capacitor current rms for the three-phase CHB inverter is based on the analysis of a single H-bridge, that of phase a. Each phase of the H-bridge has its own capacitor whose ripple current, in contrast to the two-level and NPC topologies, is not affected by the switching operations of the other phases. Hence, the analysis of a single phase H-bridge inverter is enough to provide the three-phase inverter expressions. For the CHB inverter the sum of the currents through the two upper IGBT/diode modules is used for the derivation. The calculation of the capacitor rms current is based on the analysis of one out of two symmetrical sectors (Sector I) covering the interval of $\theta = [0, \pi]$.

According to Figure 3.4 and Table 3.4:

$$i_{d,DC,I-CHB}(\theta) = \delta_a \cdot i_a \tag{3.18}$$

$$i^2_{d,rms,I-CHB}(\theta) = \delta_a \cdot i^2_a \tag{3.19}$$

The average (DC) and rms values of current i_d for the CHB inverter, are calculated using Equations 3.7 and 3.8, respectively, for $\theta = [0, \pi]$, as

$$I_{d,DC-CHB} = \frac{M I_{pk}}{2} \cos \phi \tag{3.20}$$

$$I_{d,rms-CHB} = I_{pk} \sqrt{\frac{M(3 + \cos(2\phi))}{3\pi}} \tag{3.21}$$

The rms current expression for each capacitor in this topology is then given by Equation 3.4, as

$$I_{C,rms-CHB} = I_{pk} \sqrt{\frac{M}{24\pi} [24 - 3M\pi + (8 - 3M\pi) \cos(2\phi)]} \tag{3.22}$$

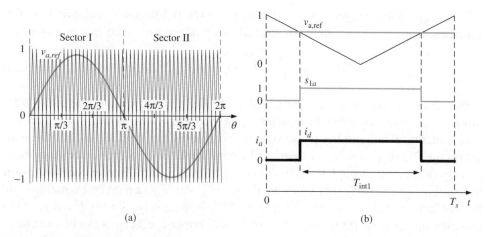

(a) $\qquad$ (b)

Figure 3.4 (a) Reference and upper carrier waveforms for the CHB inverter and (b) analysis of a switching cycle for Sector I

Table 3.4 Switching intervals for Sector I of the CHB inverter

Total duration	Total duty cycle	Current i_d
T_{int1}	δ_a	i_a

It is noted that the DC-link capacitor current of a single-phase H-bridge modulated by the SPWM strategy had been investigated in [8], deriving an expression for the rms value of HF capacitor current harmonics.

3.4 Analytical Derivation of DC-Link Capacitor Current Harmonics

In this section, the geometric wall model, a method introduced by Black in [20], is used to analytically derive the DC-link current spectra of three-level inverters. The geometric wall model provides an alternative way of representing the process of pulse generation in PWM converters. As shown in [21], the carrier and reference waveforms are redrawn in a transformed plane, so that the intersections between the new waveforms define the same train of pulses for a given PWM modulation strategy. The new plane can be divided into identical unit cells, which are characteristic for each strategy. The width of the generated pulses is periodic with respect to both dimensions of this plane, thus the function (F) that describes the pulse train can be written as a double Fourier series.

The Fourier analysis results in a spectrum that plots the function in the frequency domain. The (complex) Fourier coefficient for a harmonic, ^{mn}F, located at frequency

$nf_o + mf_c$ (n, m are integers), where f_o and f_c are the fundamental and switching frequencies, respectively, are given by the following integral over the area of a unit cell:

$$^{mn}F = \frac{1}{2\pi^2} \int_{-\pi}^{\pi} \int_{-\pi}^{\pi} F(x, y) e^{j(mx+ny)} dx dy \qquad (3.23)$$

The geometric wall model has been widely applied for analyzing the output (PWM) voltage harmonics of different inverter topologies and PWM strategies [21]. In this case, function F represents a phase voltage (commonly that of phase a). For the analysis of the DC-link capacitor current, on the other hand, F represents a module current, namely the current through the IGBT/diode module V_{1a}. The Fourier coefficients for the currents of the two other upper modules, V_{1b} and V_{1c}, are given by multiplying the coefficients for i_{V1a} by $e^{+2jn\pi/3}$ and $e^{-2jn\pi/3}$, respectively. Since current i_d is the complex sum of these three module currents, its Fourier coefficients can be calculated using the following equation:

$$^{mn}i_d = {}^{mn}i_{V1a} + {}^{mn}i_{V1b} + {}^{mn}i_{V1c} \qquad (3.24)$$

The coefficients for the DC-link capacitor current are also given by Equation 3.24, excluding the DC component ($n = m = 0$). An analysis of the DC-link capacitor current harmonics for the two-level inverter can be found in [9].

3.4.1 NPC Inverter

As in the two-level inverter, the instantaneous current flowing through the DC-link capacitor of the NPC inverter is the complex sum of the currents through the inverter's three upper modules (V_{1a}, V_{1b}, and V_{1c}), shown in Figure 3.2a. Thus, harmonic analysis of i_{V1a} is sufficient to calculate the DC-link capacitor current harmonics.

The following solution is given for SPWM implemented as the three-level phase-disposition pulse width modulation (PD PWM) [21], which uses two in-phase triangular carriers (and a sinusoidal reference waveform). Application of the geometric wall model to the three-level PD PWM results in the unit cell illustrated in Figure 3.5.

The complex Fourier coefficients of i_{V1a} come from:

$$^{mn}i_{V1a\text{-NPC}} = \frac{1}{2\pi^2} \int_{-\pi}^{\pi} \int_{-\pi}^{\pi} i_{V1a\text{-NPC}}(x, y) \, e^{j(mx+ny)} dx dy \qquad (3.25)$$

According to [21], the output voltage for phase a is positive only in the closed region (III), at the center of the graph. Hence, this is the only region where module V_{1a} carries the current of phase a. In regions I and II, the current through V_{1a} is zero. The above integral therefore reduces to:

$$^{mn}i_{V1a\text{-NPC}} = \frac{I_{pk}}{2\pi^2} \int_{-\pi/2}^{\pi/2} \int_{-\pi M \cos y}^{\pi M \cos y} \cos(y - \phi) \cdot e^{j(mx+ny)} dx dy \qquad (3.26)$$

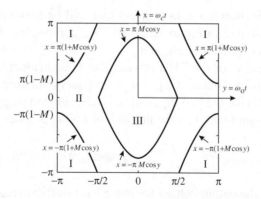

Figure 3.5 Unit cell for the three-level PD PWM method [21]

The solution of Equation 3.26 provides the complex Fourier coefficients for i_{V1a}, while Equation 3.24 gives the respective coefficients for the upper DC-link capacitor of the three-level NPC inverter as follows:

Baseband harmonics, for ($m = 0$ and) $n = 3, 9, 15, \ldots$:

$$^{0n}i_{\text{C-NPC}} = (-1)^{\frac{n-1}{2}} \frac{6MI_{\text{pk}}}{\pi(n^2 - 4)} \left(\frac{2 \cos \phi}{n} + \text{j} \sin \phi \right) \tag{3.27}$$

Carrier harmonics, for ($n = 0$ and) $m = 1, 2, 3, \ldots$:

$$^{m0}i_{\text{C-NPC}} = -\frac{3I_{\text{pk}}}{m\pi} J_1(Mm\pi) \cos \phi \tag{3.28}$$

Sideband harmonics, for n even and $m = 1, 2, 3, \ldots$:

$$^{mn}i_{\text{C-NPC}} = (-1)^{\frac{n}{2}} \frac{3I_{\text{pk}}}{2m\pi} \begin{bmatrix} \text{e}^{\text{j}\phi} J_{n-1}(Mm\pi) \\ -\text{e}^{-\text{j}\phi} J_{n+1}(Mm\pi) \end{bmatrix} \tag{3.29}$$

Sideband harmonics, for n odd and $m = 1, 2, 3, \ldots$:

$$^{mn}i_{\text{C-NPC}} = (-1)^{\frac{n-1}{2}} \frac{6I_{\text{pk}}}{m\pi^2} \sum_{k=1,3,5\ldots} J_k(Mm\pi)$$

$$\left[\frac{\cos \phi + \text{j}(n+k) \sin \phi}{1 - (n+k)^2} - \frac{\cos \phi + \text{j}(n-k) \sin \phi}{1 - (n-k)^2} \right] \tag{3.30}$$

3.4.2 CHB Inverter

As explained in Section 3.3.2, the current of each capacitor in this topology is only determined by the operation of the respective H-bridge. Hence, the derivation of a capacitor current is not given by an equation in the form of Equation 3.24.

Instead, similar to Section 3.3.2, the current that will be harmonically analyzed for this topology is i_d, shown in Figure 3.2b. The AC component of this current flows through the DC-link capacitor of phase a.

A modulation strategy for the CHB inverter that is equivalent to PD PWM for the NPC inverter is assumed. This strategy, that yields equal switching losses and the same output voltage spectra for the two topologies, is described in pages 504–506 of [21]. Pulse generation is again based on the unit cell of Figure 3.5, therefore

$$^{mn}i_{\text{d-CHB}} = \frac{1}{2\pi^2} \int_{-\pi}^{\pi} \int_{-\pi}^{\pi} i_{\text{d-CHB}}(x,y)\, e^{j(mx+ny)} dx dy \tag{3.31}$$

According to [21], the output voltage for phase a is positive in region III and negative in region(s) I, at the edges of the plot. The current i_d of the CHB inverter is therefore equal to the phase current i_a, or opposite to it $(-i_a)$ in these regions, respectively. The above integral then reduces to:

$$^{mn}i_{\text{d-CHB}} = \frac{I_{\text{pk}}}{2\pi^2} \left[\begin{array}{l} \int_{-\pi/2}^{\pi/2} \int_{-\pi M \cos y}^{\pi M \cos y} \cos(y-\phi) \cdot e^{j(mx+ny)} dx dy \\ - \int_{\pi/2}^{-\pi/2} \int_{\pi(1+M\cos y)}^{-\pi(1+M\cos y)} \cos(y-\phi) \cdot e^{j(mx+ny)} dx dy \end{array} \right] \tag{3.32}$$

which can be shown to be equal to:

$$^{mn}i_{\text{d-CHB}} = \begin{cases} \frac{I_{\text{pk}}}{\pi^2} \int_{-\pi/2}^{\pi/2} \int_{-\pi M \cos y}^{\pi M \cos y} \cos(y-\phi)\, e^{j(mx+ny)} dx dy, & \text{for } n+m \to \text{even} \\ 0, & \text{for } n+m \to \text{odd} \end{cases} \tag{3.33}$$

The solution of Equation 3.33 yields the Fourier coefficients of i_d which, apart from the DC component, are also the coefficients for the DC-link capacitor current of phase a:

Baseband harmonic, for ($m=0$ and) $n=2$, only:

$$^{02}i_{\text{C-CHB}} = \frac{MI_{\text{pk}}}{2} e^{j\phi} \tag{3.34}$$

Carrier harmonics, for ($n=0$ and) m even:

$$^{m0}i_{\text{C-CHB}} = -\frac{2I_{\text{pk}}}{m\pi} J_1(Mm\pi) \cos\phi \tag{3.35}$$

Sideband harmonics, for n even and m even:

$$^{mn}i_{\text{C-CHB}} = (-1)^{1+\frac{n}{2}} \frac{I_{\text{pk}}}{m\pi} \left[\begin{array}{l} e^{j\phi} J_{n-1}(Mm\pi) \\ -e^{-j\phi} J_{n+1}(Mm\pi) \end{array} \right] \tag{3.36}$$

Sideband harmonics, for n odd and m odd:

$$^{mn}i_{\text{C-CHB}} = (-1)^{\frac{n+1}{2}} \frac{4I_{\text{pk}}}{m\pi^2} \sum_{k=1,3,5\ldots} J_k(Mm\pi)$$

$$\left[\frac{\cos\phi + j(n+k)\sin\phi}{1-(n+k)^2} - \frac{\cos\phi + j(n-k)\sin\phi}{1-(n-k)^2} \right] \qquad (3.37)$$

3.5 Numerical Derivation of DC-Link Capacitor Current RMS Value and Voltage Ripple Amplitude

The analytical derivation of expressions for the rms value and the harmonics of the DC-link capacitor current in the previous sections assumed sinusoidal voltage reference waveforms (SPWM strategy). However, non-sinusoidal references are commonly used in practice in order to achieve higher output (line) voltages, corresponding to a maximum modulation index of $2/\sqrt{3}$ (≈ 1.1547). A numerical method for calculating the capacitor rms current in the case of such reference waveforms is proposed in this section.

The method is based on the analysis presented in Section 3.3 and operates according to the code included in Appendix B. The code divides the fundamental cycle into a number of intervals and calculates the values of Equations 3.5 and 3.6 for each of them. The cumulative sums of these values are then used to numerically evaluate Equations 3.7 and 3.8, and finally derive the rms capacitor current. Furthermore, in parallel to the rms current calculation, the proposed method calculates the amplitude of the capacitor voltage oscillations. The calculation can be performed for any modulation strategy, by accordingly changing the reference voltages (or, equivalently, the common-mode voltage) in the code. Results for the following strategies are presented in this study: (i) SPWM, (ii) SPWM with 1/6 third harmonic injection (SPWM + third harmonics), and (iii) space vector modulation (SVM) with equal distribution of duty cycles between the small vectors, implemented as a carrier-based strategy. The common-mode voltages for each strategy can be found in Appendix B.

Figure 3.6a,b plot the rms value of the capacitor current for the NPC and CHB topologies, normalized with respect to the phase rms current, I_{rms} (Norm. $I_{\text{C,rms}} = I_{\text{C,rms}}/I_{\text{rms}}$). For the NPC inverter the plot is the same for any modulation strategy (SPWM can reach up to $M = 1$), while for the CHB inverter it is shown for the SVM strategy.

The rms current rating of the DC-link capacitors is determined by the maximum value that $I_{\text{C,rms}}$ can take within the operating range of the inverter. Figure 3.6c illustrates this maximum (over the whole range of M) for each value of ϕ, for the examined modulation strategies. Given that the range of power factor (angle) is known for most inverter applications, this figure can be used to select the required capacitor rms current rating according to the applied strategy.

Similar graphs are derived below using the proposed method, for the amplitude ($\Delta V_C/2$) of the DC-link capacitor voltage ripple. Figure 3.7a,b plot this amplitude for

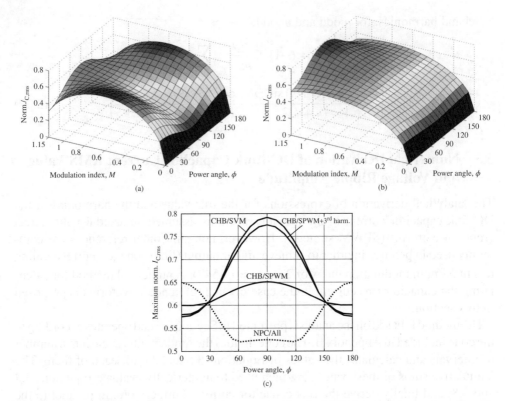

Figure 3.6 (a) Normalized rms capacitor current for the NPC inverter (same for all modulation strategies), (b) normalized rms capacitor current for the CHB inverter modulated by the SVM strategy, and (c) maximum normalized rms capacitor current for the NPC and CHB inverters, as a function of ϕ

the NPC and CHB topologies, respectively, for the case of the SVM strategy. Furthermore, Figure 3.7c,d illustrate the maximum amplitude of voltage ripple (which appears for maximum M) for the two topologies, for each modulation strategy. The presented values are normalized according to

$$\frac{\text{Norm.}\Delta V_C}{2} = \frac{f_C}{I_{rms}} \frac{\Delta V_C}{2} \tag{3.38}$$

and can be used to determine the capacitance required to limit the voltage ripple to a desired extent, $\Delta V_{C,\max}$ (Section 3.3).

3.6 Simulation Results

In this section, the derived values of the DC-link capacitor rms current, voltage ripple, and current harmonics are validated using detailed circuit simulations in MATLAB®–Simulink (SimPowerSystems toolbox). An NPC and a CHB inverter

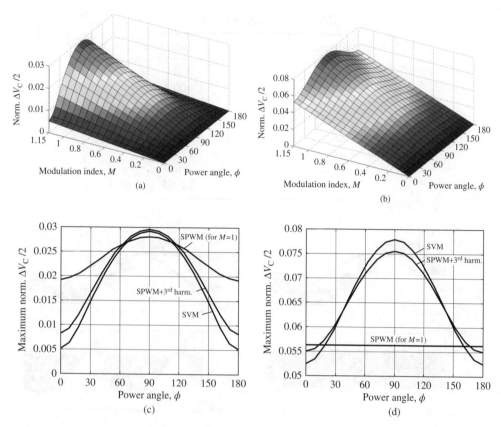

Figure 3.7 Normalized amplitude of capacitor voltage ripple for (a) the NPC and (b) the CHB inverter, modulated by the SVM strategy, and maximum normalized amplitude of capacitor voltage ripple for (c) the NPC and (d) the CHB inverter, as a function of ϕ

were simulated according to the operating parameters in Table 3.5. The capacitance value for the NPC inverter was set to 1 mF, as compared to 2.5 mF for the CHB inverter, on the basis of Equation 3.38 and Figure 3.7a,b. The estimated amplitude of voltage ripple for both topologies was then approximately 28 V, equal to a small percentage (7%) of the DC-link voltage, thus avoiding a significant distortion of the output voltages.

Figure 3.8 illustrates the line voltage v_{ab}, phase current i_a, and capacitor (C_2 for the NPC and C_a for the CHB) current i_C and voltage v_C for the two topologies. The simulated amplitude of capacitor voltage ripple agreed with the value of 28 V, calculated by the proposed method. Moreover, the rms values of the capacitor current measured during the simulations agreed with the values of 39.3 and 42.7 A given by the same method, as well as by Equations 3.17 and 3.22 for the NPC and CHB inverters, respectively.

Table 3.5 Simulation parameters

Parameter	Value
V_{DC}	400 V
I_{pk}	100 A
f_o	50 Hz
f_c	5 kHz
M	0.9
ϕ	30°
C_{NPC}	1 mF
C_{CHB}	2.5 mF

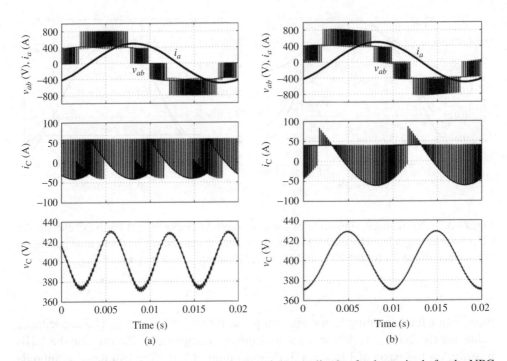

Figure 3.8 (a) DC-link capacitor rms current and (b) amplitude of voltage ripple for the NPC ($C = 1$ mF) and CHB ($C = 2.5$ mF) topologies as a function of M, for the parameters shown in Table 3.5

Figure 3.9 illustrates the capacitor current spectrum for the two topologies as derived from the above simulations, and uses it to validate the expressions presented in Section 3.4. Again, it can be seen that the frequencies and magnitudes of the current harmonics can be accurately reproduced by these expressions. Furthermore, the amplitude of the capacitor voltage ripple can be calculated by means of Equation 3.3. Considering only the baseband harmonics, this calculation gives a value of 28.6 V for both topologies, which is similar to that taken from the simulations.

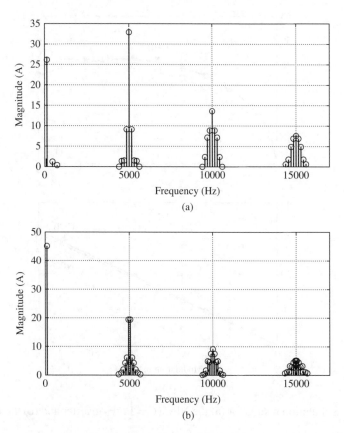

Figure 3.9 Spectrum of DC-link capacitor current for (a) the NPC and (b) the CHB inverter. Lines and circles represent the simulated and analytically derived magnitudes of the current harmonics, respectively

3.7 Discussion

3.7.1 Comparison of Capacitor Size for the NPC and CHB Inverters

The results presented in Section 3.5 were validated by simulations for a wide range of inverter operating parameters (modulation index, load power factor, fundamental, and carrier frequencies). Figure 3.10 illustrates representative results from the validation process, assuming the operating parameters in Table 3.5 and covering the whole range of M.

Figure 3.10 can also provide a comparative overview of the ripple current and capacitance requirements for the NPC and CHB topologies, at a condition ($\phi = 30°$) which is common for inverters operating as motor drives. Namely, it can be seen that the maximum ripple current is similar for both topologies, whereas the required capacitance is 2.5 times higher for the CHB inverter. Thus, each capacitor of the CHB inverter should have 2.5 times the size (equal voltage and ripple current rating, but 2.5 times more capacitance) of a capacitor for the NPC inverter.

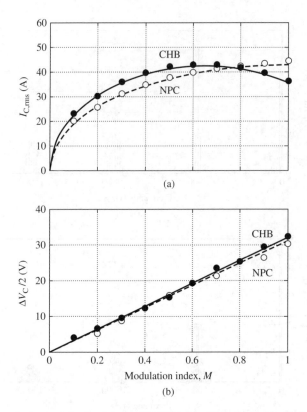

Figure 3.10 Simulation results for (a) the NPC ($C = 1$ mF) inverter and (b) the CHB inverter ($C = 2.5$ mF)

For a different value of ϕ, a similar comparison between the two topologies can be obtained by means of Figures 3.6c and 3.7c,d. As a general observation, however, we could note the following: according to Figure 3.7, for any value of ϕ, the capacitance required to obtain the same voltage ripple is at least twice that of the CHB inverter. For values of ϕ for which the ripple current is comparable for the two topologies ($\phi < 40°$, according to Figure 3.6c), each capacitor of the CHB inverter should therefore have double the size of that of the NPC inverter. Additionally, since the CHB inverter requires three (instead of two) capacitors for its DC-links, the total capacitor size will be at least three times greater for this topology. Even in cases where the ripple current is higher for the CHB inverter (ϕ approaching 90° in Figure 3.6c), a similar ratio of capacitor sizes can be expected. This is because increased capacitance (for the CHB inverter) is commonly achieved through the parallel connection of capacitors, which at the same time share the higher amounts of ripple current.

3.7.2 Comparison of Presented Methods for Analyzing DC-Link Capacitor Current

An examination of ESR characteristics of electrolytic capacitors indicates that their ESR decreases for harmonic frequencies between 50 Hz and 1 kHz, while it remains approximately constant (to $R_C = R_{C,HF}$) for higher frequencies. Inverter switching (carrier) frequencies are commonly higher than 1 kHz. As a result, the majority of carrier and sideband harmonic groups, which appear around multiples of the switching frequency, belong to the constant-ESR frequency range. Baseband harmonics, however, have to be associated with higher ESR values (Figure 3.1). Loss estimation based on rms current expressions (Section 3.3) fails to treat these harmonics separately, which results in an underestimation of DC-link electrolytic capacitor losses for three-level inverters. It is noted that this is not the case for film capacitors, since their ESR remains approximately constant throughout the current frequency range [22].

Harmonic analysis, on the other hand, can be used according to Equation 3.2 to provide a more accurate estimate for electrolytic capacitor losses. Furthermore, given that the losses arising from the carrier and sideband capacitor current harmonics can be associated with $R_{C,HF}$ in Equation 3.2, the total capacitor losses can be taken from

$$P = \sum_{B} R_C(f_h) \cdot I^2_{h,rms} + R_{C,HF} \cdot \sum_{C/S} I^2_{h,rms} \tag{3.39}$$

The subscripts B and C/S under the sums in Equation 3.39 denote that they refer to baseband and carrier/sideband harmonics, respectively.

Equation 3.39 can also be written as

$$P = \sum_{B} R_C(f_h) \cdot I^2_{h,rms} + R_{C,HF} \cdot \left[I^2_{C,rms} - \sum_{B} I^2_{h,rms} \right] \tag{3.40}$$

indicating that knowledge of the capacitor rms current and baseband harmonics is sufficient to obtain an accurate estimate for capacitor losses. The proposed numerical method can make use of Equation 3.40, based on its calculation of the capacitor current rms value and an approximation for the baseband harmonics. Namely, the baseband harmonics can be approximated by the dominant baseband harmonic, assuming that it alone generates the capacitor voltage ripple, ΔV_C (which is also calculated by the proposed strategy). Since this harmonic is located at a frequency of $3f_o$ and $2f_o$ for the NPC and CHB topologies, respectively, its amplitude is approximately given by

$$I_k = k\Delta V_C C\pi f_o \tag{3.41}$$

where k is equal to 3 and 2, accordingly. Thus, in addition to the calculation of ΔV_C, the proposed numerical method offers a more accurate estimate of capacitor losses than the analytical approach of Section 3.2. Moreover, in comparison to the harmonic analysis, it has the advantage of being easily adapted to different modulation strategies, as explained in Section 3.4. On the other hand, the harmonic analysis can provide closer estimates of capacitor power losses for inverters that operate at low switching frequencies. In these cases, certain carrier–sideband harmonic groups may belong to the LF ESR region and, therefore, the results of a complete harmonic analysis should be used according to Equation 3.2.

3.7.3 Extension to Higher-Level Inverters

This chapter focused on the DC-link capacitor current analysis of three-level NPC and CHB inverters. The presented methods can also be applied to higher-level inverters, provided that they are modulated by strategies whose reference and carrier waveforms can be defined analytically. However, such conventional modulation strategies are practically not applicable to NPC ML inverters, since they are unable to prevent voltage collapse of the intermediate DC-link capacitors. Furthermore, unconventional strategies which can retain the capacitor voltages, increase the switching frequency of the switching devices, induce higher output voltage steps (dv/dt), and distort the ML PWM waveform [2, 23].

As a consequence, ML converters with more than three-levels are predominantly based on the series connection of H-bridge cells, creating CHB ML topologies [24]. Extension to these topologies is possible for all presented methods for analyzing the DC-link capacitor current, but the complexity of the analytical derivations increases with the number of voltage levels. The proposed numerical approach, on the other hand, can be extended as shown in the relevant part of the code to cover the CHB ML topologies. As for the three-level inverter, the code simply calculates the duty cycle of each inverter cell and numerically integrates the current (and its square) through its DC-link capacitor. Results for CHB ML inverters have been validated with simulations by the authors.

3.8 Conclusion

This chapter presented a series of analytical and numerical tools for analyzing the DC-link capacitor current in three-level NPC and CHB inverters. The results can be practically useful for inverter designers performing the task of DC-link capacitor sizing. Analytical expressions were derived for the rms value and harmonics of the DC-link capacitor current, whose knowledge is essential for determining the required capacitance and ripple current rating of DC-link capacitors. Moreover, the proposed numerical method provided the possibility to estimate capacitor sizing parameters for different modulation strategies, offering a simpler alternative to repeating the analytical derivations. Results based on the proposed approach were illustrated for common modulation strategies and were validated using simulations in MATLAB®-Simulink. According to them, the CHB inverter has significantly (at least three times) higher capacitor size requirements as compared to the NPC inverter.

References

1. Rodriguez, J., Franquelo, L.G., Kouro, S. *et al.* (2009) Multilevel converters: an enabling technology for high-power applications. *Proceedings of the IEEE*, **97** (11), 1786–1817.
2. Kouro, S., Malinowski, M., Gopakumar, K. *et al* (2010) Recent advances and industrial applications of multilevel converters. *IEEE Transactions on Industrial Electronics*, **57** (8), 2553–2580.

3. Nippon Chemi-Con (2012) Aluminum Electrolytic Capacitors Catalog No. E1001M, Technical note, 2012.
4. BHC Components (2002) Aluminum Electrolytic Capacitors, Application notes, 2002.
5. Dahono, P.A., Sato, Y. and Kataoka, T. (1996) Analysis and minimization of ripple components of input current and voltage of PWM inverters. *IEEE Transactions on Industry Applications*, **32** (4), 945–950.
6. J. W. Kolar and S. D. Round, Analytical calculation of the RMS current stress on the DC-link capacitor of voltage-PWM converter systems, *IEE Proceedings - Electric Power Applications*, **153**, 4, 535-543, Jul. 2006.
7. Renken, F. (2004) Analytic calculation of the DC-link capacitor current for pulsed three-phase inverters. Proceedings of the 11th International Conference on Power Electronics and Motion Control, Riga, Latvia.
8. Renken, F. (2005) The DC-link capacitor current in pulsed single-phase H-Bridge inverters. European Conference on Power Electronics Applications, Dresden, Germany.
9. Bierhoff, M.H. and Fuchs, F.W. (2008) DC-link harmonics of three-phase voltage-source converters influenced by the pulsewidth-modulation strategy-An analysis. *IEEE Transactions on Industrial Electronics*, **55** (5), 2085–2092.
10. McGrath, B.P. and Holmes, D.H. (2009) A general analytical method for calculating inverter DC-link current harmonics. *IEEE Transactions on Industry Applications*, **45** (5), 1851–1859.
11. Nabae, A., Takahashi, I. and Akagi, H. (1981) A new neutral-point-clamped PWM inverter. *IEEE Transactions on Industry Applications*, **IA-17** (5), 518–523.
12. Zaragoza, J., Pou, J., Ceballos, S. *et al.* (2009) Voltage-balance compensator for carrier-based modulation in the neutral-point-clamped converter. *IEEE Transactions on Industrial Electronics*, **56** (2), 305–314.
13. Pou, J., Pindado, R., Boroyevich, D. and Rodriguez, P. (2005) Evaluation of the low-frequency neutral-point voltage oscillations in the three-level inverter. *IEEE Transactions on Industrial Electronics*, **52** (6), 1582–1588.
14. Gupta, K. and Khambadkone, M. (2007) A simple space vector PWM scheme to operate a three-level NPC inverter at high modulation index including overmodulation region, with neutral point balancing. *IEEE Transactions on Industry Applications*, **43** (3), 751–760.
15. Wang, C. and Li, Y. (2010) Analysis and calculation of zero-sequence voltage considering neutral-point potential balancing in three-level NPC converters. *IEEE Transactions on Industrial Electronics*, **57** (7), 2262–2271.
16. Pou, J., Zaragoza, J., Ceballos, S. *et al.* (2012) A Carrier-Based PWM strategy with zero-sequence voltage injection for a three-level neutral-point-clamped converter. *IEEE Transactions on Power Electronics*, **27** (2), 642–651.
17. Soto, D. and Green, T.C. (2002) A comparison of high-power converter topologies for the implementation of FACTS controllers. *IEEE Transactions on Industrial Electronics*, **49** (5), 1072–1080.
18. Orfanoudakis, G.I., Sharkh, S.M., Yuratich, M.A., and Abu-Sara, M.A. (2010) Loss comparison of two and three-level inverter topologies. 5th IET International Conference on Power Electronics, Machines and Drives (PEMD).
19. Orfanoudakis, G.I., Sharkh, S.M., Yuratich, M.A., and Abu-Sara, M.A. (2010) Analysis of DC-Link capacitor losses in three-level neutral point clamped and cascaded H-Bridge voltage source inverters. IEEE International Symposium on Industrial Electronics (ISIE).

20. Black, H.S. (1953) *Modulation Theory*, Van Nostrand, New York.
21. Holmes, D.G. and Lipo, T.A. (2003) *Pulse Width Modulation for Power Converters*, IEEE Press Series on Power Engineering, IEEE Press, Piscataway, NJ.
22. Wen, H., Xiao, W., Wen, X. and Armstrong, P. (2012) Analysis and evaluation of DC-link capacitors for high-power-density electric vehicle drive systems. *IEEE Transactions on Vehicular Technology*, **61** (7), 2950–2964.
23. Saeedifard, M., Iravani, R. and Pou, J. (2007) Analysis and control of DC-capacitor-voltage-drift phenomenon of a passive front-end five-level converter. *IEEE Transactions on Industrial Electronics*, **54** (6), 3255–3266.
24. Malinowski, M., Gopakumar, K., Rodriguez, J. and Pérez, M.A. (2010) A survey on cascaded multilevel inverters. *IEEE Transactions on Industrial Electronics*, **57** (7), 2197–2206.

4

Loss Comparison of Two- and Three-Level Inverter Topologies

4.1 Introduction

The process of selecting the topology, components, and operating parameters (voltage, current, and switching frequency) of an inverter is highly affected by the anticipated inverter losses. An accurate estimate of the losses occurring in each part of an inverter can significantly contribute to achieving an enhanced inverter design. This chapter examines the semiconductor and DC-link capacitor losses of four voltage source inverter topologies: the conventional two-level inverter, the two-level two-channel interleaved inverter, the three-level neutral-point-clamped (NPC) inverter, and the three-level cascaded H-bridge (CHB) inverter, shown in Figure 4.1.

Losses in two-level inverters have been reported extensively in the literature. Researchers have also investigated semiconductor losses in three-level inverters. Estimates of switching losses have been obtained using approximations of IGBT and diode $I-V$ switching characteristics [1–4]. However, a more convenient approach based on calculating switching loss using the switching energy–current ($E_{sw}-I$) characteristics, reveals that the switching losses of an IGBT–diode pair are approximately proportional to the switching voltage and current [5]. This observation can be verified based on IGBT–diode module data sheets [6]. Analytical expressions for switching losses in a two-level inverter can be found in [5, 7]. For the two-level inverter, all continuous PWM (pulse width modulation) methods have the same switching losses, which are also independent of the load phase angle [5, 8]. Discontinuous strategies, however, can result in lower switching losses. Switching losses in a three-level NPC inverter have been investigated in [7], using a second order approximation of the IGBT and diode $E_{sw}-I$ characteristics.

Expressions for the two-level inverter conduction losses can be found in [2, 5, 9]. The calculation of conduction losses is based on the linear $I-V$ characteristics of the IGBT–diode modules. Unlike switching losses, two-level inverter conduction

Power Electronic Converters for Microgrids, First Edition. Suleiman M. Sharkh, Mohammad A. Abusara, Georgios I. Orfanoudakis and Babar Hussain.
© 2014 John Wiley & Sons, Ltd. Published 2014 by John Wiley & Sons, Ltd.
Companion Website: www.wiley.com/go/sharkh

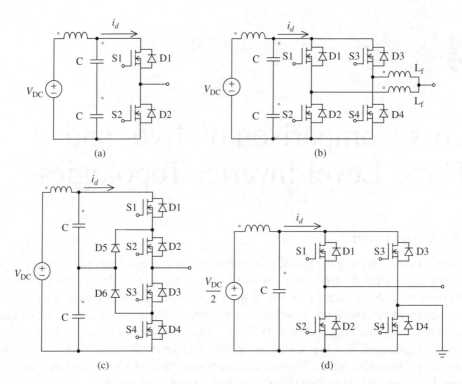

Figure 4.1 Circuit diagrams (one leg) of (a) two-level inverter, (b) two-channel two-level interleaved inverter, (c) three-level NPC inverter, and (d) three-level cascaded H-bridge inverter

losses are affected by the selection of the PWM strategy and the load power factor. Expressions for the NPC inverter can be found in [7] as well as in [10] for a number of modulation strategies.

DC-link capacitor loss estimation is based on the rms value of the capacitor current. The derivation of the current rms expression for the two-level inverter has been presented in [6, 11, 12]. Capacitor loss estimations also appear for the two-level two-channel interleaved and the three-level CHB inverters in [13, 14], respectively. The DC-link of the three-level NPC inverter has only been studied in the literature with respect to its voltage and the capacitor balancing problem [15–17].

In this chapter, expressions for switching and conduction losses in the four inverter topologies are reviewed. Analytical expression for DC-link capacitor losses are derived for the two-level interleaved and the three-level inverters. Unlike most studies that focus on a single inverter topology or loss type, the expressions for semiconductor and DC-link capacitor losses are used to compare the four examined topologies. The comparison is based on a selection of inverter components from available commercial devices with appropriate voltage/current ratings and switching frequency. This selection, which is different for each topology, affects the resulting losses.

4.2 Selection of IGBT-Diode Modules

The four inverter topologies are compared on the basis of a common power output. Assuming a DC-link voltage V_{DC} of 2 kV (1 kV for the CHB inverter) and a nominal load peak current I_M of 370 A, the inverter power rating S_o is equal to 555 kVA.

The switching voltage of the IGBT–diode modules in a three-level inverter is half of that in a two-level inverter generating a voltage waveform with the same amplitude. The voltage rating of the IGBT–diode modules used in a three-level inverter therefore needs to be half that of an equivalent two-level inverter. This difference in voltage rating has a very significant impact on switching and conduction loss parameters of the modules.

The current carried by each module is the same for all topologies except for the interleaved inverter in which each module carries half the current. The effect of the module current rating on switching loss parameters is insignificant, but conduction loss parameters are approximately doubled for the half current-rated modules.

Appropriate IGBT modules are selected for each topology. The two-level inverter uses high-voltage high-current IGBT–diode modules (A), the interleaved inverter uses high-voltage low-current modules (B), while the three-level inverters use the low-voltage high-current module (C). Table 4.1 lists the switching and conduction parameters of the selected modules. Module A is the Eupec FZ800R33KL2C_B5 3.3 kV, 800 A IGBT–diode module, Module B is the FZ400R33KL2C_B5 3.3 kV, 400 A IGBT–diode module, while Module C is the FF800R17KE3 1.7 kV, 800 A module. Parameter values have been obtained from the modules' data sheets [6].

Parameters $V_{0,c/d}$ and $R_{c/d}$ approximate the conduction I–V characteristics of IGBTs/diodes, respectively, according to:

$$V = V_{0,c/d} + I \cdot R_{c/d} \tag{4.1}$$

Table 4.1 IGBT–diode module parameters

Parameter	Module A	Module B	Module C	Unit
V_{base}	1.8	1.8	0.9	kV
$V_{0,c}$	1.6	1.6	0.9	V
R_c	2.5	5	1.87	mΩ
$V_{0,d}$	1.7	1.7	1	V
R_d	1.25	2.5	1	mΩ
a_c	5.7	5.7	0.8	mJ A^{-1}
b_c	50	50	40	mJ
a_d	0.5	0.5	0.12	mJ A^{-1}
b_d	150	150	60	mJ

Parameters V_{base}, $a_{c/d}$, and $b_{c/d}$ approximate the switching energy E_{sw}–I characteristics according to:

$$E_{1sw,c/d} = \frac{V_{sw}}{V_{base}}(a_{c/d} \cdot I_{sw} + b_{c/d}) \tag{4.2}$$

where V_{sw} and I_{sw} are the instantaneous switching voltage and current, respectively. The equations are written in a compact form, to cover both IGBTs (subscript "c") and diodes (subscript "d"), see also Equation 4.3.

4.3 Switching Losses

For a given switching frequency f_{sw}, the two-level inverter has the same switching losses for all continuous PWM methods. Switching losses are also independent of the inverter modulation index M and the load power factor PF [5] but increase linearly with switching frequency. Conduction losses are not affected by f_{sw} but depend on the modulation strategy, M and PF. For commonly used switching frequencies, conduction losses of the two-level inverter are significantly lower than corresponding switching losses.

The two-level two-channel interleaved inverter losses are examined under the assumption that the instantaneous current carried by each of the inverter channels is approximately half of the respective phase current. Leg (channel) inductors and sufficiently high switching frequencies are used to satisfy this requirement. Each module in this topology therefore carries half the current of a two-level inverter module. On the other hand, the number of modules in the interleaved inverter is twice that of the conventional two-level inverter.

4.3.1 Switching Losses in the Two-Level Inverters

Figure 4.2 illustrates the turn-on and turn-off energy losses of an IGBT with respect to its switching current. Similar graphs are also available for power diodes. The sum of the two curves for each value of the switching current represents the total switching energy loss during one complete switching. The curves are approximately linear and hence this sum can be approximated by a first order equation with respect to the instantaneous switching current, I_{sw}. Switching losses are also proportional to the switching voltage, V_{sw}, and therefore the switching energy loss characteristics (i.e., the energy losses during one on–off switching), $E_{1sw,c}$ for IGBTs and $E_{1sw,d}$ for diodes, can be mathematically described by equations of the following form:

$$\left. \begin{array}{l} E_{1sw,c} = \frac{V_{sw}}{V_{base}}\left(a_c \cdot I_{sw} + b_c\right) \\ E_{1sw,d} = \frac{V_{sw}}{V_{base}}(a_d \cdot I_{sw} + b_d) \end{array} \right\} \text{ or } E_{1sw,c/d} = \frac{V_{sw}}{V_{base}}(a_{c/d} \cdot I_{sw} + b_{c/d}) \tag{4.3}$$

The latter equation is written in a compact form to cover both IGBTs (subscript c) and diodes (subscript d). The curve-fitting constants a_c and b_c are used for the IGBTs,

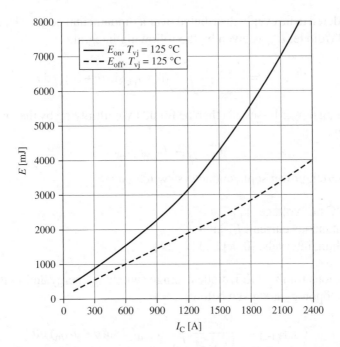

Figure 4.2 IGBT turn-on and turn-off energy loss characteristics

while a_d and b_d are used for the diodes. Symbols a and b represent the sums $a_c + a_d$ and $b_c + b_d$, respectively. V_{base} represents the reference switching voltage, used for the derivation of the plot in the data sheet.

In addition to the switching energy characteristics, the calculation of average switching losses for a given inverter semiconductor, requires the following parameters:

- Switching frequency, f_{sw}.
- Switching voltage (i.e., the voltage across which the semiconductor has to switch), V_{sw}.
- Switching current (i.e., the current that flows/will flow through the semiconductor, before/after it switches off/on, respectively). For reasons of simplicity of the derived expressions, angle θ in this section will represent the current angle instead of the voltage reference angle. Thus, the switching current will be $I_{pk}\sin\theta$.
- Angle interval(s) during which the semiconductor switches, expressed as portion(s) of the fundamental cycle, $[\theta_1, \theta_2]$.

The first step for the derivation of a switching loss expression is the calculation of the average energy loss, E_{sw}, over a fundamental period. One complete switching happens during each switching period. Under the assumption that the switching frequency is several times higher than the fundamental frequency, the switching events can be considered to be evenly distributed over a fundamental period. Thus, for a given inverter IGBT/diode, E_{sw} is given by the following integral:

$$E_{sw,c/d} = \frac{V_{sw}}{V_{base}} \frac{1}{2\pi} \int_{\theta_1}^{\theta_2} (a_{c/d} I_{pk} \sin \theta + b_{c/d}) \, d\theta \tag{4.4}$$

The switching power losses can then be found by multiplying by the inverter switching frequency:

$$P_{sw,c/d} = E_{sw,c/d} f_{sw} \tag{4.5}$$

In a 2L inverter, power semiconductors switch under:

- The full DC bus voltage: V_{DC}
- The instantaneous current: $I_{pk} \sin\theta$
- The switching intervals: $[0, \pi]$ and $[\pi, 2\pi]$.

The expressions for the IGBT/diode average switching energy and switching power losses in this topology are given below:

$$E_{sw,c/d\text{-}2L} = \frac{V_{DC}}{V_{base}} \frac{1}{2\pi} \int_0^\pi (a_{c/d} I_{pk} \sin \theta + b_{c/d}) \, d\theta$$

$$P_{sw,c/d\text{-}2L} = \frac{V_{DC}}{V_{base}} \left(\frac{a_{c/d}}{\pi} I_{pk} + \frac{b_{c/d}}{2} \right) f_{sw} \tag{4.6}$$

According to Equation 4.6, the power loss expression is derived from the energy loss expression in a simple manner, and hence, throughout the chapter, the two expressions will be presented as a pair with a single equation number. The expression is the same for IGBTs and diodes. Since the three-phase 2L inverter consists of six IGBTs–diodes, its total switching losses will be given by

$$P_{sw\text{-}2L,3\phi} = 6 \frac{V_{DC}}{V_{base}} \left(\frac{a}{\pi} I_{pk} + \frac{b}{2} \right) f_{sw} \tag{4.7}$$

where

$$a = a_c + a_d$$
$$b = b_c + b_d \tag{4.8}$$

The switching losses in the interleaved inverter losses can be shown to be given by:

$$P_{sw,2L\text{-}Int} = 6 \frac{V_{DC}}{V_{base}} \left(\frac{a}{\pi} I_{pk} + b \right) f_{sw} \tag{4.9}$$

4.3.2 Switching Losses in the NPC Inverter

In an NPC 3L inverter, IGBTs and diodes switch under:

- Half the DC bus voltage: $V_{DC}/2$
- Instantaneous current: $I_{pk} \sin \theta$
- Switching intervals:
  - $(S_1) \leftrightarrow (D_5)$ switch during $[0, \pi - \phi]$
  - $(S_2, D_5) \leftrightarrow (D_3, D_4)$ switch during $[\pi - \phi, \pi]$
  - $(S_4) \leftrightarrow (D_6)$ switch during $[\pi, 2\pi - \phi]$
  - $(S_3, D_6) \leftrightarrow (D_1, D_2)$ switch during $[2\pi - \phi, 2\pi]$.

The average switching energy and the respective power loss expressions for each semiconductor of this topology will be ("/" is used as in Equations 4.1–4.3 to provide a compact equation form for semiconductors that share the same loss expression):

$$E_{sw\text{-}NPC,S1/D5} = E_{sw\text{-}NPC,S4/D6} = \frac{V_{DC}}{V_{base}} \frac{1}{4\pi} \int_0^{\pi-\phi} (a_{c/d} I_{pk} \sin \theta + b_{c/d}) d\theta$$

$$P_{sw\text{-}NPC,S1/D5} = P_{sw\text{-}NPC,S4/D6} = \frac{V_{DC}}{V_{base}} \frac{1}{4\pi} [a_{c/d} I_{pk}(\cos \phi + 1) + b_{c/d}(\pi - \phi)] f_{sw}$$

$$(4.10)$$

$$E_{sw\text{-}NPC,S2/S3} = \frac{V_{DC}}{V_{base}} \frac{1}{4\pi} \int_{\pi-\phi}^{\pi} (a_c I_{pk} \sin \theta + b_c) d\theta$$

$$P_{sw\text{-}NPC,S2/S3} = \frac{V_{DC}}{V_{base}} \frac{1}{4\pi} [a_c I_{pk}(1 - \cos \phi) + b_c \phi] f_{sw} \qquad (4.11)$$

$$E_{sw\text{-}NPC,D1/\ldots/D4} = \frac{V_{DC}}{V_{base}} \frac{1}{8\pi} \int_{\pi-\phi}^{\pi} (a_d I_{pk} \sin \theta + b_d) d\theta$$

$$P_{sw\text{-}NPC,D1/\ldots/D4} = \frac{V_{DC}}{V_{base}} \frac{1}{8\pi} [a_d I_{pk}(1 - \cos \phi) + b_d \phi] f_{sw} \qquad (4.12)$$

According to the above expressions, the total switching losses for the three-phase 3L NPC inverter are:

$$P_{sw\text{-}NPC,3\phi} = 3 \frac{V_{DC}}{V_{base}} \left(\frac{a}{\pi} I_{pk} + \frac{b}{2} \right) f_{sw} \qquad (4.13)$$

4.3.3 Switching Losses in the CHB Inverter

In a CHB 3L inverter, the semiconductors switch under conditions which are similar to those for the NPC inverter:

- Half the DC bus voltage: $V_{DC}/2$
- Instantaneous current: $I_{pk} \sin\theta$
- Switching intervals:
  - $(S_1) \leftrightarrow (D_2)$ switch during $[0, \pi - \phi]$
  - $(S_4) \leftrightarrow (D_3)$ switch during $[\pi - \phi, \pi]$
  - $(S_3) \leftrightarrow (D_4)$ switch during $[\pi, 2\pi - \phi]$
  - $(S_2) \leftrightarrow (D_1)$ switch during $[2\pi - \phi, 2\pi]$.

The switching loss expressions for the IGBTs $S_1 \ldots S_4$ and diodes $D_1 \ldots D_4$ for each H-bridge of the CHB topology, will be the following:

$$E_{\text{sw-CHB},S1/D2} = E_{\text{sw-CHB},S3/D4} = \frac{V_{DC}}{V_{\text{base}}} \frac{1}{4\pi} \int_0^{\pi-\phi} (a_{c/d} I_{pk} \sin\theta + b_{c/d}) d\theta$$

$$P_{\text{sw-CHB},S1/D2} = P_{\text{sw-CHB},S3/D4} = \frac{V_{DC}}{V_{\text{base}}} \frac{1}{4\pi} [a_{c/d} I_{pk}(\cos\phi + 1) + b_{c/d}(\pi - \phi)] f_{\text{sw}}$$

$$(4.14)$$

$$E_{\text{sw-CHB},S4/D3} = E_{\text{sw-CHB},S4/D3} = \frac{V_{DC}}{V_{\text{base}}} \frac{1}{4\pi} \int_{\pi-\phi}^{\pi} (a_{c/d} I_{pk} \sin\theta + b_{c/d}) d\theta$$

$$P_{\text{sw-CHB},S4/D3} = P_{\text{sw-CHB},S4/D3} = \frac{V_{DC}}{V_{\text{base}}} \frac{1}{4\pi} [a_{c/d} I_{pk}(1 - \cos\phi) + b_{c/d}\phi] f_{\text{sw}} \quad (4.15)$$

The total switching losses for the three-phase CHB comes from the following expression, which is identical to the respective one for the NPC inverter:

$$P_{\text{sw-CHB},3\phi} = 3\frac{V_{DC}}{V_{\text{base}}} \left(\frac{a}{\pi} I_{pk} + \frac{b}{2}\right) f_{\text{sw}} = P_{\text{sw-NPC},3\phi} \quad (4.16)$$

4.4 Conduction Losses

Conduction losses occur during the semiconductor conduction intervals, due to their non-zero on-state voltage drops. Two main approaches have been used in the literature for the estimation of conduction losses, both giving the same results. The first [2] assumes a high switching frequency and divides the fundamental period into infinitesimal time intervals, equal to the switching period. Then, it multiplies the expressions for the power loss and the duty cycle of each power semiconductor for a given time instant with the elementary switching period (dt). Finally, it integrates this product over a fundamental period and obtains the average conduction losses. The second approach [9] performs a pulse-by-pulse analysis that examines the exact positions and durations of each semiconductor's conduction intervals. It does not assume an infinite switching frequency, hence approaching what happens in an inverter more closely, but yields complicated intermediate results. The final results of the two methods are identical, which verifies the applicability of the first approach even in lower switching frequencies.

Unlike switching losses, inverter conduction losses are affected by the selection of the PWM strategy and the load power factor. Expressions for the 2L inverter conduction losses can be found in [2, 5, 9]. Regarding the NPC inverter, expressions for IGBT and diode conduction losses are derived in [5, 7, 18], and have been verified in the context of the present work. Expressions for the NPC inverter can also be found in [1, 10] for a number of modulation strategies. No literature has been found on the conduction losses of the CHB and the interleaved inverter topologies.

IGBT and diode conduction (instead of switching) I–V characteristics are used for deriving expressions for conduction losses. In order to approximate these characteristics, it is common to use two curve-fitting constants, $V_{0,c/d}$ and $R_{c/d}$ that represent the semiconductor's saturation voltage and on-state resistance, respectively. Their values are acquired from I–V graphs included in IGBT–diode module data sheets.

Figure 4.3 [6] presents the conduction I–V characteristic for an IGBT. $V_{0,c}$ is equal to the voltage at the point where the extension of the linear part of the graph intersects the horizontal axis, while R_c is equal to the inverse of its gradient. $V_{0,c/d}$ and $R_{c/d}$ approximate the conduction I–V characteristics of IGBTs–diodes, according to

$$V = V_{0,c/d} + I\,R_{c/d} \tag{4.17}$$

The derivation of the conduction losses for an inverter semiconductor, associates its conduction characteristics with the following parameters:

- The instantaneous load current amplitude, $I_{pk}\sin\theta$ and power factor, $PF = \cos\phi$
- The conduction interval(s) during a fundamental period, $[\theta_1, \theta_2]$
- The conduction duty cycle as a function of the voltage angle, $\delta(\theta + \phi)$.

The term "conduction interval" in the present context is used macroscopically to denote the portions (in terms of current angle θ) of the fundamental period during which the examined IGBT or diode carries current while it is turned on. The device does not conduct during the whole conduction interval but during a fraction of it, which is equal to the instantaneous duty cycle. Under the assumption of a high inverter switching frequency, the sum of all elementary conduction losses over a fundamental period can be approximated by a respective integral. Division by the duration of a

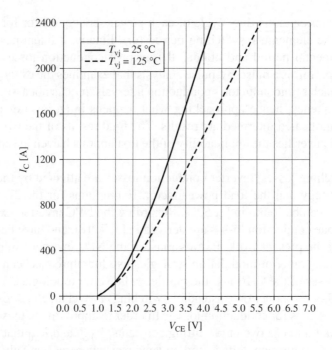

Figure 4.3   IGBT I–V characteristic

fundamental cycle (2π) gives the expression for conduction power losses of a specific IGBT/diode of the inverter:

$$P_{con} = \frac{1}{2\pi} \int_{\theta_1}^{\theta_2} \delta(\theta + \phi) I_{c/d} V_{c/d} d\theta$$

or

$$P_{con} = \frac{1}{2\pi} \int_{\theta_1}^{\theta_2} \delta(\theta + \phi) I_{pk} \sin \theta (R_{c/d} I_{pk} \sin \theta + V_{0,c/d}) d\theta \qquad (4.18)$$

4.4.1   Conduction Losses in the Two-Level Inverter

In a 2L inverter, power semiconductors conduct:

- The instantaneous load current: $I_{pk} \sin\theta$
- Conduction intervals:
  - IGBT S_1 and diode D_2 conduct during $[0, \pi]$
  - IGBT S_2 and diode D_1 conduct during $[\pi, 2\pi]$
- Duty cycles:
  - IGBTs: $\delta = \frac{1}{2}[1 + M \sin(\theta + \phi)]$
  - Diodes: $\delta = \frac{1}{2}[1 - M \sin(\theta + \phi)]$.

According to the previous paragraph, the IGBT, diode, and total conduction losses for the three-phase 2L inverter will be given by the following expressions:

$$P_{\text{con-2L,S1/S2}} = \frac{1}{2\pi} \int_0^\pi \frac{1}{2}[1 + M\sin(\theta + \phi)]I_{\text{pk}}\sin\theta(R_c I_{\text{pk}}\sin\theta + V_{0,c})d\theta$$

$$P_{\text{con-2L,S1/S2}} = \left(\frac{1}{8} + \frac{M}{3\pi}\cos\phi\right)R_c I_{\text{pk}}^2 + \left(\frac{1}{2\pi} + \frac{M}{8}\cos\phi\right)V_{0,c}I_{\text{pk}} \tag{4.19}$$

$$P_{\text{con-2L,D1/D2}} = \frac{1}{2\pi} \int_0^\pi \frac{1}{2}[1 - M\sin(\theta + \phi)]I_{\text{pk}}\sin\theta(R_d I_{\text{pk}}\sin\theta + V_{0,d})d\theta$$

$$P_{\text{con-2L,D1/D2}} = \left(\frac{1}{8} - \frac{M}{3\pi}\cos\phi\right)R_d I_{\text{pk}}^2 + \left(\frac{1}{2\pi} - \frac{M}{8}\cos\phi\right)V_{0,d}I_{\text{pk}} \tag{4.20}$$

$$P_{\text{con-2L,3}\phi} = \left[\frac{3}{4}(R_c + R_d) + \frac{2M}{\pi}(R_c - R_d)\cos\phi\right]I_{\text{pk}}^2$$
$$+ \left[\frac{3}{\pi}(V_{0,c} + V_{0,d}) + \frac{3M}{4}(V_{0,c} - V_{0,d})\cos\phi\right]I_{\text{pk}} \tag{4.21}$$

The conduction losses in the interleaved inverter can be shown to be given by:

$$P_{\text{con,2L-Int}} = \left[\frac{3}{8}(R_c + R_d) + \frac{M}{\pi}(R_c - R_d)\cos\phi\right]I_{\text{pk}}^2$$
$$+ \left[\frac{3}{\pi}(V_{c0} + V_{d0}) + \frac{3M}{4}(V_{c0} - V_{d0})\cos\phi\right]I_{\text{pk}} \tag{4.22}$$

4.4.2 Conduction Losses in the NPC Inverter

As for the 2L inverter, the power semiconductors that conduct in this topology conduct the instantaneous load current. The conduction intervals for each semiconductor of the NPC inverter are summarized in Table 4.2. The voltage waveform produced by 3L inverters switches between $\pm V_{\text{DC}}/2$ and 0, not between $+V_{\text{DC}}/2$ and $-V_{\text{DC}}/2$ as in 2L inverters. This affects the calculation of the semiconductor duty cycles, which are different than the 2L inverter's and are shown in Table 4.3. Note that in this topology certain devices conduct during two non-adjacent intervals during a fundamental cycle.

The resulting conduction power loss expressions for the inverter IGBTs are the following:

$$P_{\text{con-NPC,S1/S4}} = \frac{1}{2\pi} \int_0^{\pi-\phi} M\sin(\theta + \phi)I_{\text{pk}}\sin\theta(R_c I_{\text{pk}}\sin\theta + V_{0,c})d\theta$$

$$P_{\text{con-NPC,S1/S4}} = \frac{MR_c I_{\text{pk}}^2}{6\pi}(1 + \cos\phi)^2 + \frac{MV_{0,c}I_{\text{pk}}}{4\pi}[(\pi - \phi)\cos\phi + \sin\phi] \tag{4.23}$$

Table 4.2　Conduction intervals for the NPC inverter

Device conducting	From	To	With duty cycle		From	To	With duty cycle
S_1	0	$\pi - \phi$	δ_1	AND			
S_2	0	$\pi - \phi$	1		$\pi - \phi$	π	δ_3
D_3, D_4	$\pi - \phi$	π	δ_4				
D_5	0	$\pi - \phi$	δ_2		$\pi - \phi$	π	δ_3
S_4	π	$2\pi - \phi$	δ_4				
S_3	π	$2\pi - \phi$	1		$2\pi - \phi$	2π	δ_2
D_1, D_2	$2\pi - \phi$	2π	δ_1				
D_6	π	$2\pi - \phi$	δ_3		$2\pi - \phi$	2π	δ_2

Table 4.3　Conduction duty cycles for three-level inverters

Sign of $v_{x,\mathrm{ref}}$	Leg x in state	With duty cycle	Symbol
$v_{x,\mathrm{ref}} \geq 0$	(P)	$M\sin(\theta + \phi)$	δ_1
	(Z)	$1 - M\sin(\theta + \phi)$	δ_2
$v_{x,\mathrm{ref}} < 0$	(Z)	$1 + M\sin(\theta + \phi)$	δ_3
	(N)	$-M\sin(\theta + \phi)$	δ_4

$$P_{\mathrm{con\text{-}NPC},S2/S3} = \frac{1}{2\pi}\left[\int_0^{\pi-\phi} I_{\mathrm{L}} \sin\theta (R_{\mathrm{c}}I_{\mathrm{pk}} \sin\theta + V_{0,\mathrm{c}})\mathrm{d}\theta \right.$$

$$\left. + \int_{\pi-\phi}^{\pi} [1 + M\sin(\theta + \phi)]I_{\mathrm{pk}} \sin\theta (R_{\mathrm{c}}I_{\mathrm{pk}} \sin\theta + V_{0,\mathrm{c}})\mathrm{d}\theta\right] \quad (4.24)$$

$$P_{\mathrm{con\text{-}NPC},S2/S3} = \frac{R_{\mathrm{c}}I_{\mathrm{pk}}^2}{4} - \frac{R_{\mathrm{c}}I_{\mathrm{pk}}^2 M}{6\pi}(1 - \cos\phi)^2$$

$$+ \frac{V_{0,\mathrm{c}}I_{\mathrm{pk}}}{\pi} + \frac{MV_{0,\mathrm{c}}I_{\mathrm{pk}}}{4\pi}(\phi\cos\phi - \sin\phi)$$

The respective conduction power loss expressions for the inverter diodes are:

$$P_{\mathrm{con\text{-}NPC},D1/\ldots/D4} = \frac{1}{2\pi}\int_{\pi-\phi}^{\pi} [-M\sin(\theta + \phi)]I_{\mathrm{pk}} \sin\theta (R_{\mathrm{d}}I_{\mathrm{pk}} \sin\theta + V_{0,\mathrm{d}})\mathrm{d}\theta$$

$$P_{\mathit{con-NPC},D1/\ldots/D4} = \frac{MR_{\mathrm{d}}I_{\mathrm{pk}}^2}{6\pi}(1 - \cos\phi)^2 + \frac{MV_{0,\mathrm{d}}I_{\mathrm{pk}}}{4\pi}(\sin\phi - \phi\cos\phi) \quad (4.25)$$

$$P_{\text{con-NPC,D5/D6}} = \frac{1}{2\pi} \left[\int_0^{\pi-\phi} [1 - M\sin(\theta + \phi)] I_{\text{pk}} \sin\theta (R_d I_{\text{pk}} \sin\theta + V_{0,d}) d\theta \right.$$

$$\left. + \int_{\pi-\phi}^{\pi} [1 + M\sin(\theta + \phi)] \dots d\theta \right] \qquad (4.26)$$

$$P_{\text{con-NPC,D5/D6}} = \frac{R_d I_{\text{pk}}^2}{4} - \frac{M R_d I_{\text{pk}}^2}{3\pi}(1 + \cos^2\phi)$$

$$+ \frac{V_{0,d} I_{\text{pk}}}{\pi} + \frac{M V_{0,d} I_{\text{pk}}}{4\pi}[(2\phi - \pi)\cos\phi - 2\sin\phi]$$

Using the above expressions, the total conduction losses for the three-phase NPC inverter can be proved to be the following:

$$P_{\text{con-NPC},3\phi} = \left[\frac{3}{2}(R_c + R_d) + \frac{4M}{\pi}(R_c - R_d)\cos\phi \right] I_{\text{pk}}^2$$

$$+ \left[\frac{6}{\pi}(V_{0,c} + V_{0,d}) + \frac{3M}{2}(V_{0,c} - V_{0,d})\cos\phi \right] I_{\text{pk}} \qquad (4.27)$$

4.4.3   Conduction Losses in the CHB Inverter

Instantaneous current values and duty cycle definitions are identical for the 3L NPC and CHB inverters. However, the conduction intervals, illustrated in Table 4.4, are different for the CHB topology.

Table 4.4   Conduction intervals for the CHB inverter

Device conducting	From	To	With duty cycle		From	To	With duty cycle
S_1	0	$\pi - \phi$	δ_1	AND			
S_2	π	$2\pi - \phi$	1		$2\pi - \phi$	2π	δ_2
D_1	$2\pi - \phi$	2π	δ_1				
D_2	0	$\pi - \phi$	δ_2		$\pi - \phi$	π	1
S_3	π	$2\pi - \phi$	δ_4				
S_4	0	$\pi - \phi$	1		$\pi - \phi$	π	δ_3
D_3	$\pi - \phi$	π	δ_4				
D_4	π	$2\pi - \phi$	δ_3		$2\pi - \phi$	2π	1

The resulting conduction power loss expressions for the CHB inverter's IGBTs and diodes are given below:

$$P_{\text{con-CHB,S1/S3}} = \frac{1}{2\pi} \int_0^{\pi-\phi} [M \sin(\theta + \phi)] I_{\text{pk}} \sin \theta (R_c I_{\text{pk}} \sin \theta + V_{0,c}) d\theta$$

$$P_{\text{con-CHB,S1/S3}} = \frac{MR_c I_{\text{pk}}^2}{6\pi}(1 + \cos \phi)^2 + \frac{MV_{0,c}I_{\text{pk}}}{4\pi}[(\pi - \phi)\cos \phi + \sin \phi] \quad (4.28)$$

$$P_{\text{con-CHB,S2/S4}} = \frac{1}{2\pi} \left[\int_0^{\pi-\phi} I_{\text{pk}} \sin \theta (R_c I_{\text{pk}} \sin \theta + V_{0,c}) d\theta \right.$$

$$\left. + \int_{\pi-\phi}^{\pi} [1 + M \sin(\theta + \phi)] I_{\text{pk}} \sin \theta (R_c I_{\text{pk}} \sin \theta + V_{0,c}) d\theta \right]$$

$$P_{\text{con-CHB,S2/S4}} = \frac{R_c I_{\text{pk}}^2}{4} - \frac{MR_c I_{\text{pk}}^2}{6\pi}(1 - \cos \phi)^2$$

$$+ \frac{V_{0,c}I_{\text{pk}}}{\pi} + \frac{MV_{0,c}I_{\text{pk}}}{4\pi}(\phi \cos \phi - \sin \phi) \quad (4.29)$$

$$P_{\text{con-CHB,D1/D3}} = \frac{1}{2\pi} \int_{\pi-\phi}^{\pi} [-M \sin(\theta + \phi)] I_{\text{pk}} \sin \theta (R_d I_{\text{pk}} \sin \theta + V_{0,d}) d\theta$$

$$P_{\text{con-CHB,D1/D3}} = \frac{MR_d I_{\text{pk}}^2}{6\pi}(1 - \cos \phi)^2 + \frac{MV_{0,d}I_{\text{pk}}}{4\pi}(\sin \phi - \phi \cos \phi) \quad (4.30)$$

$$P_{\text{con-CHB,D2/D4}} = \frac{1}{2\pi} \int_0^{\pi-\phi} [1 - M \sin(\theta + \phi)] I_{\text{pk}} \sin \theta (R_c I_{\text{pk}} \sin \theta + V_{0,c}) d\theta$$

$$+ \frac{1}{2\pi} \int_{\pi-\phi}^{\pi} I_{\text{pk}} \sin \theta (R_c I_{\text{pk}} \sin \theta + V_{0,c}) d\theta$$

$$P_{\text{con-CHB,D2/D4}} = \left[-\frac{M}{6\pi}(1 + \cos \phi)^2 + \frac{1}{4} \right] R_d I_{\text{pk}}^2$$

$$+ \left[\frac{M}{4\pi}[(\phi - \pi)\cos \phi - \sin \phi] + \frac{1}{\pi} \right] I_{\text{pk}} V_{0,d} \quad (4.31)$$

As for the switching losses, the expression for the total conduction losses of the three-phase CHB inverter can be shown to be identical to the respective expression for the NPC inverter (Equation 4.27):

$$P_{\text{con-CHB},3\phi} = P_{\text{con-NPC},3\phi} \quad (4.32)$$

4.5   DC-Link Capacitor RMS Current

In this section, the method of [19], used for the derivation of the two-level inverter capacitor current rms expression, is applied to the three other inverter topologies. See also Chapter 3 for detailed discussion and analysis of capacitor current in the three-level topologies. The method considers each inverter IGBT–diode module as a switch that, while on, carries the current of the respective phase. The sum of the currents through the upper switches of an inverter is i_d, as shown in Figure 4.1 for each of the four topologies. The DC component of this current is supplied by the inverter DC source, while the AC component is filtered, and hence carried by the DC-link capacitor. The rms value of the capacitor current, $I_{C,rms}$, is calculated using the average (DC) and rms values of i_d, $I_{d,DC}$, and $I_{d,rms}$, respectively:

$$I_{d,rms}^2 = I_{d,DC}^2 + I_{C,rms}^2 \Rightarrow I_{C,rms} = \sqrt{I_{d,rms}^2 - I_{d,DC}^2} \qquad (4.33)$$

According to [19], the calculation of the current i_d average and rms values is based on the analysis of its transitions during a single switching period. If i_d is equal to $i_{d,int1}$, $i_{d,int2}$, ... during time intervals T_{int1}, T_{int2}, ..., respectively (with $T_{intk} < T_s$), then its average and rms values during a switching period T_s are given by the following equations:

$$i_{d,DC}(\theta) = \frac{1}{T_s} \left(\sum_k T_{int\,k} \cdot i_{d,int\,k} \right) = \sum_k \delta_{int\,k} \cdot i_{d,int\,k} \qquad (4.34)$$

$$i_{d,rms}^2(\theta) = \frac{1}{T_s} \left(\sum_k T_{int\,k} \cdot i_{d,int\,k}^2 \right) = \sum_k \delta_{int\,k} \cdot i_{d,int\,k}^2 \qquad (4.35)$$

The interval duty cycles $\delta_{int\,k}$ and respective currents $i_{d,int\,k}$ are functions of θ, the angle of the voltage reference waveform for phase a. The average and rms values of i_d over a fundamental period are obtained using the following expressions:

$$I_{d,DC} = \frac{1}{2\pi} \int_0^{2\pi} i_{d,DC}(\theta) d\theta \qquad (4.36)$$

$$I_{d,rms} = \sqrt{\frac{1}{2\pi} \int_0^{2\pi} i_{d,rms}^2(\theta) d\theta} \qquad (4.37)$$

The expressions for $\delta_{intk}(\theta)$ and $i_{d,intk}(\theta)$ may change during sectors of the fundamental cycle. In this case, the above expressions are written as sums of integrals for the different sectors.

For example, for the derivation of the DC-link capacitor current rms expression of the two-level two-channel interleaved inverter, the fundamental cycle is divided into

Table 4.5 Switching intervals for the two-level two-channel interleaved inverter

	Sector I_1	Sector I_2
Duration (θ)	$0-\pi/6$	$\pi/6-\pi/3$
T_a/T_s	δ_a	δ_a
T_b/T_s	δ_b	δ_b
T_c/T_s	δ_c	δ_c
Interval 1	$2(1-\delta_c-\delta_b)\rightarrow -i_b/2$	$2(1-\delta_a-\delta_b)\rightarrow -i_b/2$
Interval 2	$2(\delta_c-\delta_a)\rightarrow (i_c-i_b)/2$	$2(\delta_a-\delta_c)\rightarrow (i_a-i_b)/2$
Interval 3	$2(\delta_a-1/2)\rightarrow -i_b$	$2(\delta_c-1/2)\rightarrow -i_b$

six sectors, each of which covers an angle of $\pi/3$. The inverter operation in these sectors is symmetric and hence only one sector needs to be analyzed. Sector I $(0-\pi/3)$ is divided into two sub-sectors, as described in Table 4.5. The table illustrates the duty cycles of the switching intervals and the corresponding values of current i_d as $\delta_{intk}\rightarrow i_{intk}$. Angles θ_a, θ_b, and θ_c, are equal to:

$$\theta_a = 0, \theta_b = \frac{2\pi}{3} \quad \text{and} \quad \theta_c = -\frac{2\pi}{3}$$

For two-level inverters:

$$\delta_x = \frac{1}{2}(1 + M\sin(\theta + \theta_x)) \tag{4.38}$$

Assuming that the inverter load has a power factor $\cos(\phi)$, the three-phase currents are given by:

$$i_x = I_M \sin(\theta + \theta_x - \phi) \tag{4.39}$$

According to Table 4.5, for sector I_1:

$$i_{d,DC,I1,2L\text{-}Int}(\theta) = 2\cdot(1-\delta_c-\delta_b)\cdot\left(\frac{-i_b}{2}\right) + 2\cdot(\delta_c-\delta_a)\cdot\left(\frac{i_c-i_b}{2}\right) + 2\cdot(\delta_a-1/2)\cdot(-i_b) \tag{4.40}$$

$$i^2_{d,rms,I1,2L\text{-}Int}(\theta) = 2\cdot(1-\delta_c-\delta_b)\cdot\left(\frac{-i_b}{2}\right)^2 2\cdot(\delta_c-\delta_a)\cdot\left(\frac{i_c-i_b}{2}\right)^2 + 2\cdot(\delta_a-1/2)\cdot(-i_b)^2 \tag{4.41}$$

Similarly, expressions can be derived for sector I_2. The average (DC) and rms values of i_d are given by:

$$I_{d,DC,2L\text{-}Int} = \frac{3}{\pi}\left(\int_0^{\pi/6} i_{d,DC,I1,2L\text{-}Int}d\theta + \int_{\pi/6}^{\pi/3} i_{d,DC,I2,2L\text{-}Int}d\theta\right) \tag{4.42}$$

$$I_{d,rms,2L\text{-}Int} = \sqrt{\frac{3}{\pi}\left(\int_0^{\pi/6} i^2_{d,rms,I1,2L\text{-}Int}d\theta + \int_{\pi/6}^{\pi/3} i^2_{d,rms,I2,2L\text{-}Int}d\theta\right)} \tag{4.43}$$

which results in:

$$I_{\text{d,DC,2L-Int}} = MI_M \frac{\pi}{4} \cos(\phi) \tag{4.44}$$

$$I_{\text{d,rms,2L-Int}} = I_M \sqrt{\frac{3M}{\pi} \left[\frac{\sqrt{3}-1}{24} + \frac{\sqrt{3}+2}{6} \cos^2(\phi) \right]} \tag{4.45}$$

The capacitor current rms expression for the two-level interleaved inverter will therefore be given by:

$$I_{\text{C,rms,2L-Int}} = I_M \sqrt{\frac{3M}{\pi} \left[\frac{\sqrt{3}-1}{24} + \frac{\sqrt{3}+2}{6} \cos^2(\phi) \right] - \frac{\pi^2 M^2 \cos^2(\phi)}{16}} \tag{4.46}$$

For the three-level NPC inverter, the three-phase voltage and current waveforms are divided into three sectors, covering an interval of $2\pi/3$, each. Sector I is divided into three sub-sectors, as described in Table 4.6.

For three-level inverters:

$$\delta_x = M \sin(\theta + \theta_x) \tag{4.47}$$

According to Table 4.5, for sector I_1:

$$i_{\text{d,DC,I1,NPC}}(\theta) = \delta_c \cdot (-i_b) + (\delta_a - \delta_c) \cdot i_a \tag{4.48}$$

$$i^2_{\text{d,rms,I1,NPC}}(\theta) = \delta_c \cdot (-i_b)^2 + (\delta_a - \delta_c) \cdot i^2_a \tag{4.49}$$

Similar expressions are derived for sectors I_2 and I_3. The DC and rms values of i_d for the NPC inverter are derived using equations similar to Equations 4.44 and 4.45, which results in:

$$I_{\text{d,DC,NPC}} = \frac{3}{4} MI_M \cos\phi \tag{4.50}$$

$$I_{\text{d,rms,NPC}} = I_M \sqrt{\frac{\sqrt{3}M}{4\pi}[1 + \cos^2(\phi)]} \tag{4.51}$$

Table 4.6 Switching intervals for the three-level NPC inverter

	Sector I_1	Sector I_2	Sector I_3
Duration (θ)	$\pi/6-\pi/3$	$\pi/3-2\pi/3$	$2\pi/3-5\pi/6$
T_a/T_s	δ_a	δ_a	δ_a
T_b/T_s	0	0	δ_b
T_c/T_s	δ_c	0	0
Interval 1	$\delta_c \rightarrow -i_b$	$\delta_a \rightarrow i_a$	$\delta_b \rightarrow -i_c$
Interval 2	$\delta_a - \delta_c \rightarrow i_a$		$\delta_a - \delta_b \rightarrow i_a$

Use of Equation 4.33 gives the rms current expression for the upper capacitor of the three-level NPC inverter. Due to symmetry, the expression for the lower capacitor is identical:

$$I_{C,rms,NPC} = I_M \sqrt{\frac{M}{2} \left[\frac{\sqrt{3}}{2\pi} + \left(\frac{2\sqrt{3}}{\pi} - \frac{9}{8}M \right) \cos^2(\phi) \right]} \tag{4.52}$$

The derivation of the DC-link capacitor current rms for the three-phase CHB inverter is based on the analysis of a single H-bridge, that of phase a. The current rms value of each capacitor in this topology is not affected by the switching operations of the other phases. The calculation of the capacitor rms current is based on the analysis of one out of two symmetrical sectors, covering an interval of π, each.

According to Table 4.7:

$$i_{d,DC,Casc}(\theta) = \delta_a \cdot i_a \tag{4.53}$$

$$i^2_{d,rms,Casc}(\theta) = \delta_a \cdot i^2_a \tag{4.54}$$

The average (DC) and rms values of current i_d for the CHB inverter are calculated using Equations 4.36 and 4.37, respectively, which results in:

$$I_{d,DC,Casc} = \frac{MI_M}{2} \cos(\phi) \tag{4.55}$$

$$I_{d,rms,Casc} = I_M \sqrt{\frac{M(3 + \cos(2\phi))}{3\pi}} \tag{4.56}$$

The capacitor current rms expression for this topology can be calculated using Equations 4.33, 4.55, and 4.56 to be:

$$I_{C,rms,Casc} = I_M \sqrt{\frac{M}{24\pi}[24 - 3M\pi + (8 - 3M\pi)\cos(2\phi)]} \tag{4.57}$$

The DC-link capacitor current of a single-phase H-bridge has been investigated in [14], deriving an expression for the rms value of high frequency capacitor current harmonics. This expression is equivalent to Equation 4.57, which also incorporates the low-frequency harmonics of the CHB inverter DC-link capacitor current. All the above derived capacitor current rms expressions were verified by inverter simulations in the SimPowerSystems toolbox of MATLAB®-Simulink.

Table 4.7   Switching intervals for phase a of the three-level cascaded H-bridge inverter

Duration (θ)	$0-\pi$
T_a/T_s	δ_a
Interval 1	$\delta_a \to i_a$

4.6 Results

All inverters are assumed to supply a 3Ω impedance (Z) load with power factor equal to 0.9. Due to the increased switching losses of high-voltage IGBTs, however, two-level inverters are assumed to be switched at lower frequencies. The switching frequency f_s is set to 1 kHz for two-level and 2.5 kHz for three-level inverters, respectively. Figure 4.4 plots the semiconductor losses against the inverter modulation index M, according to the equations in Section 4.4 and the values in Table 4.1. An inspection of the plot indicates that the switching losses of the two-level inverters are significantly higher than those of the three-level inverters. Even though the switching frequencies of the two-level inverters are lower, three-level inverters exhibit a major advantage over switching losses, as a result of their decreased switching parameter values ($a_{c/d}$ and $b_{c/d}$). The decreased number of output voltage levels and the lower switching frequencies also have a negative impact on the output harmonic performance of the two-level inverters, but this consideration is beyond the scope of this chapter.

The conduction losses of the two two-level inverters are equal due to the values of conduction parameters $R_{c/d}$, which are half for the interleaved inverter modules. Conduction losses for three-level topologies are slightly higher.

The DC-link capacitors power losses are given by the following expression:

$$P_C = N \cdot R_C \cdot I_{C,rms}^2 \qquad (4.58)$$

where N is the number of capacitors used in each topology and R_C represents the equivalent series resistance (ESR) of each capacitor. As shown in Figure 4.1, N is equal to 2 for the two two-level and the three-level NPC inverters, and 3 for the CHB inverter. Each capacitor (or capacitor bank) is assumed to have a nominal voltage of 1 kV and an ESR of 15 mΩ.

Figure 4.5 illustrates the variation of total DC-link capacitor losses with modulation index for the examined topologies. As shown in the figure, the two-level interleaved

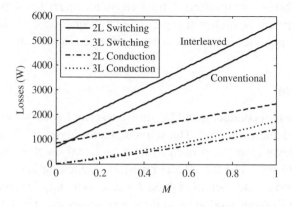

Figure 4.4 Semiconductor losses versus M

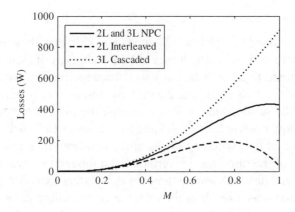

Figure 4.5   DC-link capacitor losses versus M

inverter has the lowest amount of DC-link capacitors losses. The capacitor losses in the conventional two-level inverter are higher and equal to the three-level NPC inverter losses. In fact, the expressions for the capacitor current rms values of these two topologies are identical. However, according to inverter simulations, their instantaneous capacitor currents and current spectra differ significantly. In contrast to the two-level inverter, the capacitor current of the three-level NPC inverter contains low-frequency harmonics. Low-frequency harmonics also appear in the capacitor current of the three-level CHB inverter. This topology has the highest amount of capacitor losses, partially because it uses three instead of two DC-link capacitors.

A fixed value of ESR was assumed for all DC-link capacitors. In reality, the ESR of electrolytic capacitors that are commonly used for inverter DC-links varies with the frequency of capacitor current harmonics. Its value for low frequencies, in the range of hundreds of hertz, is two to three times higher than it is for frequencies in the range of kilohertz.

Losses for the three-level inverters, whose capacitors carry low-frequency capacitor currents, will therefore be higher than predicted by Equation 4.58. DC-link capacitor sizing for these topologies must consider this expected increment.

4.7   Conclusion

In this chapter the semiconductor and DC-link capacitor losses of four inverter topologies are examined and compared. The semiconductor losses of the conventional and interleaved two-level inverters proved to be significantly higher than the respective losses of the NPC and CHB three-level inverters. Switching losses that dominate in the two-level inverters are increased even for low switching frequencies, due to the high-voltage IGBT–diode modules that these topologies use. The interleaved inverter has higher switching losses than the conventional two-level inverter. Semiconductor losses for the three-level topologies are lower and equal for the NPC and CHB inverters, assuming an equivalence of their modulation strategies.

In terms of DC-link capacitor losses, the interleaved inverter can achieve better results than the conventional two-level inverter. Capacitor losses in the conventional two-level and three-level NPC inverters proved to be equal. The CHB inverter, on the contrary, has significantly more capacitor losses than these two topologies. Lower DC-link capacitor losses of the two-level inverters cannot compensate for their increased semiconductor losses. Given the similarity of three-level inverter semiconductor losses, the NPC inverter proved to be the most efficient of the four topologies.

References

1. Al-Naseem, O., Erickson, R.W., and Carlin, P. (2000) Prediction of switching loss variations by averaged switch modelling. IEEE Applied Power Electronics Conference (APEC), Vol. 1, pp. 242–248.
2. Casanellas, F. (1994) Losses in PWM inverters using IGBTs. *IEE Proceedings in Electrical Power Applications*, **141** (5), 235–239.
3. Rajapakse, A.D., Gole, A.M. and Wilson, P.L. (2005) Electromagnetic transients simulation models for accurate representation of switching losses and thermal performance in power electronic systems. *IEEE Transactions on Power Delivery*, **20** (1), 319–327.
4. Rajapakse, A.D., Gole, A.M., and Wilson, P.L. (2005) Approximate loss formulae for estimation of IGBT switching losses through EMTP-type simulations. International Conference on Power Systems Transients (IPST).
5. Kolar, J.W., Ertl, H. and Zach, F.C. (1999) Influence of the modulation method on the conduction and switching losses of a PWM inverter system. *IEEE Transactions on Industry Applications*, **27** (6), 1063–1075.
6. Eupec Infineon IGBT/Diode Module Data Sheets. www.infineon.com/eupec. (accessed January 2010).
7. Dieckerhoff, S., Bernet, S. and Krug, D. (2005) Power loss-oriented evaluation of high voltage IGBTs and multilevel converters in transformerless traction applications. *IEEE Transactions on Power Electronics*, **20** (6), 1379.
8. Hava, A.M., Kerkman, R.J. and Lipo, T.A. (1999) Simple analytical and graphical methods for carrier-based PWM-VSI drives. *IEEE Transactions on Power Electronics*, **14** (1), 49–61.
9. Mestha, L.K. and Evans, P.D. (1989) Analysis of on-state losses in PWM inverters. *IEE Proceedings*, **136** (4), 189–195.
10. Wang, Q., Chen, Q., Jiang, W., and Hu, C. (2007) Analysis and comparison of conduction losses in neutral-point-clamped three-level inverter with PWM control. International Conference on Electrical Machines and Systems (ICEMS), Seoul, South Korea.
11. Kolar, J.W., Wolbank, T.M., and M. Schrodl (1999) Analytical calculation of the RMS current stress on the DC link capacitor of voltage DC link PWM converter systems. IEE 9th International Conference on Electrical Machines and Drives, No 468.
12. Renken, F. (2004) Analytic calculation of the dc-link capacitor current for pulsed three phase inverters. EPE-PEMC.

13. Asiminoaei, L., Aeloiza, E., Enjeti, P.N. and Blaabjerg, F. (2008) Shunt active-power-filter topology based on parallel interleaved inverters. *IEEE Transactions on Industrial Electronics*, **55** (3), 1175–1189.
14. Renken, F. (2005) The dc-link capacitor current in pulsed single-phase H-bridge inverters. EPE European Conference on Power Electronics and Applications.
15. Celanovic, N. and Boroyevich, D. (2000) A comprehensive study of neutral-point voltage balancing problem in three-level neutral-point-clamped voltage source PWM inverters. *IEEE Transactions on Power Electronics*, **15** (2), 242–249.
16. Munduate, A., Garin, I., Figueres, E., and Garcera, G. (2006) Analytical study of the DC link capacitors voltage ripple in three level neutral point clamped inverters. International Symposium on Power Electronics, Electrical Drives, Automation and Motion (SPEEDAM).
17. Ogasawara, S. and Akagi, H. (1993) Analysis of variation of neutral point potential in neutral-point-clamped voltage source PWM inverters. IEEE Industry Applications Society (IAS) Annual Meeting.
18. Kim, T.J., Kong, D.W., Lee, Y.H., and Hyun, D.S. (2001)The analysis of conduction and switching losses in multilevel-inverter system. IEEE Power Electronics Specialists Conference (PESC), Vol. **3**, pp. 1363–1368.
19. Dahono, P.A., Sato, Y. and Kataoka, T. (1996) Analysis and Minimization of ripple components of input current and voltage of PWM inverters. *IEEE Transactions on Power Electronics*, **32** (4), 945–950.

5

Minimization of Low-Frequency Neutral-Point Voltage Oscillations in NPC Converters

5.1 Introduction

The three-level neutral-point-clamped (NPC) converter (Figure 5.1), invented three decades ago [1], is currently the most widely used topology in industrial medium-voltage applications [2]. Its advantage over the traditional two-level converter is expanding its range of application to low-voltage systems, such as photovoltaic (PV inverters) and low-voltage motor drives [3]. Major power module manufacturers have recently started producing three-level NPC modules to meet the increasing demand [4, 5]. Compared to the two-level topology, the NPC converter is built using modules with half the voltage rating, which decreases the total amount of switching losses [3], as discussed in Chapter 4. The converter can therefore operate at higher switching frequencies and achieve an increased efficiency, while generating a lower-THD (total harmonic distortion) (three-level) PWM (pulse width modulation) phase voltage waveform. On the other hand, the NPC topology has the disadvantages of higher switch and driver count, as well as having to cope with DC-link capacitor imbalance [6].

The DC-link of the NPC converter consists of two capacitors, C_1 and C_2, connected at the converter neutral point (NP), shown in Figure 5.1. The design and operation of the converter are based on the assumption that the voltages of the two DC-link capacitors, v_{C1} and v_{C2}, respectively, are kept approximately balanced, that is, equal to $V_{DC}/2$, each. Equivalently, the NP voltage, v_{NP}, defined as

$$v_{NP} = v_{C1} - V_{DC}/2 \qquad (5.1)$$

in this study (Figure 5.1), is assumed to be kept approximately equal to zero. Capacitor balancing is an essential requirement for the NPC converter, since voltage imbalance beyond a certain extent causes voltage stress on the overcharged capacitor, as well as on the converter modules that switch across it. Excessive imbalance that appears

Power Electronic Converters for Microgrids, First Edition. Suleiman M. Sharkh, Mohammad A. Abusara, Georgios I. Orfanoudakis and Babar Hussain.
© 2014 John Wiley & Sons, Ltd. Published 2014 by John Wiley & Sons, Ltd.
Companion Website: www.wiley.com/go/sharkh

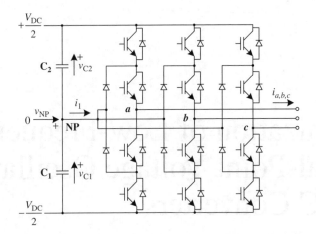

Figure 5.1   Three-level NPC converter

recurrently during the converter operation also increases the chances for premature module failure due to cosmic radiation [7]. In addition, the imbalance distorts the output (PWM) voltage waveform, unless it is compensated by the converter modulation strategy [8].

Voltage imbalance can appear in two forms during the operation of the NPC converter: (i) as a voltage deviation, typically appearing during a transient, and (ii) as a low-frequency voltage oscillation during the converter's steady-state operation. The first type of imbalance will be referred to as transient imbalance, while the low-frequency DC-link capacitor voltage oscillation is widely known as NP voltage ripple. A system front end providing two separate DC sources of $V_{DC}/2$, each, connected at the NP, can be a solution to the balancing problem. However, in the typical case when the NPC converter is supplied by a single DC source, such as a 12- or 18-pulse rectifier, the task of DC-link capacitor balancing, or simply NP balancing, has to be solely carried out by the converter modulation strategy.

This chapter proposes new concepts for the modulation of the NPC converter, according to which a family of modulation strategies can be created. These strategies offer the advantage of generating the minimum possible NP voltage ripple, without significantly affecting the converter's rated switching frequency or spectral performance. The following section places the proposed strategies in the context of the state-of-the-art modulation strategies for the NPC converter, providing further details for the contribution of this study.

5.2 NPC Converter Modulation Strategies

A modulation strategy for the NPC converter should be capable of bringing the capacitors back to balance after a transient imbalance, and minimizing the NP voltage ripple during steady-state operation. Several strategies have been proposed to fulfill

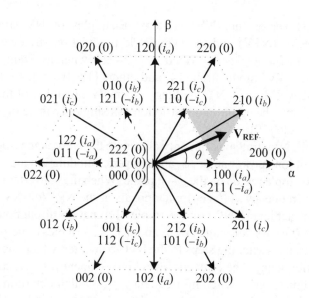

Figure 5.2   Space vector diagram for a three-level NPC converter. The NP current corresponding to each vector is shown in parentheses

these requirements, using either a space vector [9–16] or a carrier-based implementation [17–23]. The analysis presented in this study uses the terms of space vector modulation (SVM). However, it encompasses carrier-based strategies, since there is a well-known equivalence between the two ways of implementation [21, 24, 25].

The space vector diagram for a three-level NPC converter with balanced DC-link capacitors is illustrated in Figure 5.2. The triplets used as vector names denote the states (0 for $-V_{DC}/2$, 1 for v_{NP}, or 2 for $+V_{DC}/2$) of phase voltages a, b, and c, respectively. An SVM strategy selects a number of vectors, $\mathbf{V}_1$, $\mathbf{V}_2$, ..., $\mathbf{V}_n$ and adjusts their duty cycles $d_1, d_2, \ldots, d_n$, respectively, to create the reference vector, $\mathbf{V}_{REF}$ (see Section 2.2.2):

$$\mathbf{V}_{REF} = d_1\mathbf{V}_1 + d_2\mathbf{V}_2 + \ldots + d_n\mathbf{V}_n \qquad (5.2)$$

$$d_1 + d_2 + \ldots + d_n = 1 \qquad (5.3)$$

Assuming that the converter does not operate in the over-modulation region, then the reference vector falls in one of the small triangles defined by the dashed lines. The vectors at the vertices of this triangle form a set that can be used, according to Equations 5.2 and 5.3, to create $\mathbf{V}_{REF}$. For a given $\mathbf{V}_{REF}$, these are also the "nearest vectors" (NVs) on the SV plane. In Figure 5.2, for example, $\mathbf{V}_{REF}$ falls in the shaded triangle, so the NVs are: 210, 221, 110, 211, and 100.

The modulation strategies for the NPC converter can then be divided into two categories. A strategy can be characterized as an NV strategy if, in order to form the reference vector, it is only allowed to select among the respective NVs [26]. Otherwise, if it has the freedom to use additional vectors, it can be characterized as a non-nearest vector (non-NV) strategy. Examples of NV strategies are the nearest-three-vector (NTV) and the symmetric modulation strategies, described in [10], as well as the strategies described in [17–21], whereas examples of non-NV strategies are the NTV^2 modulation [14] and others [15, 16, 23]. At this point, it can be noted that the well-known NTV strategy is simply a member of the family of NV strategies, which has the additional restriction of using only three of the NVs during each switching cycle [10].

As shown in [6], NV strategies can only eliminate the NP voltage ripple for a certain range of values of load power angle, ϕ, and converter modulation index, m. These values define a zero-ripple region on the $\phi-m$ plane. When the converter operates outside this region, low-frequency voltage ripple appears at the NP. Non-NV strategies have the advantage of achieving NP voltage ripple elimination throughout the converter operating range (i.e., for all values of ϕ and m). Nevertheless, the use of non-NVs introduces additional switching steps to the NPC converter, which increases its effective switching frequency. The effective switching frequency, $f_{\text{sw,eff}}$, can be defined in addition to the converter's switching frequency, f_{sw} (which determines the duration, T_{sw}, of the switching cycle), and can be used in relation to switching losses:

$$f_{\text{sw,eff}} = f_{\text{sw}} \cdot \frac{\text{average number of switching steps per } T_{\text{sw}}}{6} \quad (5.4)$$

Conventionally, when the converter is modulated by a continuous carrier-based strategy, each of its three legs switches exactly twice (rising–falling or vice versa) during every switching period. Hence, six switching steps take place in total during T_{sw}, and therefore $f_{\text{sw,eff}} = f_{\text{sw}}$. By appropriately defining their switching sequences, NV strategies can also operate with an average of six switching steps per switching cycle, and thus achieve the above value of $f_{\text{sw,eff}}$. Furthermore, less steps per cycle, and thus lower effective switching frequencies can be achieved by NV (SVM) strategies that correspond to discontinuous carrier-based strategies, which, however, produce lower-quality output voltage waveforms [10, 21]. In non-NV strategies, on the other hand, two more switching steps are added per cycle, therefore $f_{\text{sw,eff}}$ rises to $4f_{\text{sw}}/3$ [14, 15]. This 33.3% increase has a notable effect on the converter switching losses, which is the main drawback of non-NV strategies. Additionally, during each switching period, non-NVs cause one of the converter legs to switch between $+V_{\text{DC}}/2$, (possibly) v_{NP}, and $-V_{\text{DC}}/2$. This generates phase voltage pulses similar to those of a two-level converter, therefore distorting the standard three-level PWM phase voltage waveform and increasing its harmonic distortion (weighted total harmonic distortion, WTHD) [11, 14–16, 23].

Both NV and non-NV strategies are used in practice for the modulation of the three-level NPC converter, at the expense of NP voltage ripple, or increased switching losses and output voltage WTHD, respectively. This trade-off has also led to the creation of hybrid strategies, which operate as combinations of NV and non-NV strategies. Hybrid strategies mitigate the drawbacks of non-NV strategies, while limiting the NP voltage ripple to a value that can be tolerated by the converter

[11, 27, 28]. Nevertheless, increased cooling and filtering requirements, as well as decreased converter efficiency, still arise from the use of hybrid (due to the participation of non-NV) strategies. NV strategies that avoid these disadvantages while abiding by the tolerable voltage limits would therefore be desirable.

This chapter continues in Section 5.3 with an analytical derivation of a lower boundary (approximately equivalent to a minimum) for the NP voltage ripple that can be achieved by an NV strategy. In Section 5.4, it describes a concept for the operation of NV strategies, which provides the capability of achieving this minimum. Then, in Section 5.5, it examines the performance of the NV strategies created according to this concept, showing that they can decrease the NP voltage ripple of existing NV strategies by up to 50%, depending on the NPC converter's operating point on the ϕ–m plane. Section 5.6 verifies the results by simulations in MATLAB®-Simulink. The practical significance and applicability of the whole approach are discussed in Section 5.7. Finally, Section 5.8 discusses and proposes hybrid modulation strategies for eliminating NP voltage ripple.

5.3   Minimum NP Ripple Achievable by NV Strategies

The space vector diagram for a three-level converter can be divided into six 60° sextants, for voltage reference angle, $\theta = [0, 60)°$, $[60, 120)°$, and so on. In each sextant, there are two long vectors (**L0** and **L1**), one medium vector (**M**), four small vectors (**S0$_1$**, **S0$_2$**, **S1$_1$**, and **S1$_2$**), and three zero vectors (**Z$_1$**, **Z$_2$**, and **Z$_3$**), which are arranged as shown in Figure 5.3.

The operation of an NV modulation strategy is summarized in the following steps, which are repeated in each switching cycle:

Step 1 – Determination of NVs: According to **V**$_{REF}$ (see Figure 5.2).
Step 2 – Calculation of duty cycles: The duty cycles of the nearest vectors are derived from Equations 5.2 and 5.3. However, since the two small vectors of each pair share the same position on the SV plane, solution of Equation 5.2 gives the

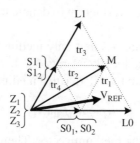

Figure 5.3   Space vectors and triangles tr$_1$–tr$_4$, in one sextant of a three-level converter. By convention, **S0$_1$** and **S1$_1$** stand for positive small vectors (i.e., corresponding to i_a, i_b, or i_c), whereas **S0$_2$** and **S1$_2$** stand for negative small vectors in Figure 5.2

duty cycles d_{S0} and d_{S1} as totals that can be distributed among $\mathbf{S0_1}$–$\mathbf{S0_2}$ and $\mathbf{S1_1}$–$\mathbf{S1_2}$, respectively (see Step 3 below). Similarly, the duty cycle for zero vectors, d_Z, is calculated as a total that can be distributed among vectors $\mathbf{Z_1}$, $\mathbf{Z_2}$, and $\mathbf{Z_3}$.

Step 3 – Distribution of duty cycles: d_{S0} and d_{S1} are distributed among the two vectors of the respective pairs, according to a criterion that aims to minimize the NP voltage ripple. Additionally, NV strategies commonly impose a number of constraints on this distribution (switching constraints), to reduce the converter effective switching frequency. Switching constraints also determine the distribution of d_Z among the three zero vectors.

Step 4 – Selection of switching sequence: Each NV strategy has a predefined set of switching sequences, formed in accordance with the strategy's switching constraints. Depending on the results from the previous steps, one of these sequences is selected to successively activate the vectors according to the assigned duty cycles.

The following analysis is based on these four steps, to derive a lower boundary for the NP voltage ripple that can be achieved by an NV strategy.

5.3.1 Locally Averaged NP Current

Let us assume a three-level NPC converter operating at a steady-state, having a balanced set of sinusoidal output currents with rms value I_o, which *lead* the phase voltages by the power angle, ϕ. The converter is modulated with an NV strategy at a modulation index, m. The reference frequency is f, while the switching frequency is f_{sw}. The dc-link voltage is V_{DC} and the capacitors C_1 and C_2 have a capacitance C, each.

The NP voltage ripple that appears at the DC-link is generated by the NP current, i_1, shown in Figure 5.1. Each of the space vectors in Figure 5.2 can be related to a specific NP current, shown in parentheses. This current is equal to the sum of the currents of the phases that are at state "1", since these phases are connected to the NP. For example, the NP current corresponding to vector 110 is $-i_c$, because $i_a + i_b = -i_c$. The function $I_{NP}(\mathbf{V})$ will be used further on to denote the value of the NP current that corresponds to vector $\mathbf{V}$.

It can be observed from Figure 5.2 that only medium and small vectors produce a non-zero NP current. If i_M is the locally averaged (i.e., the average during a switching period) current that is taken from the NP due to a medium vector, $\mathbf{M}$ (Figure 5.3), then

$$i_M = d_M I_{NP}(\mathbf{M}) \tag{5.5}$$

where d_M is the duty cycle of the medium vector. The value of i_M is a function of the angle θ. Since the selection and duty cycles of medium vectors are solely determined by Equations 5.2 and 5.3, the waveform of i_M is the same for any NV modulation strategy.

On the other hand, if i_S is the locally averaged NP current that can be taken from the NP due to small vectors, then i_S depends on the distribution of d_{S0} and d_{S1} among the small vectors of the respective pairs. This is because the two small vectors of each pair produce opposite values of NP current (Figure 5.2). Two distribution factors, x_{S0} and x_{S1} ($x_{S0}, x_{S1} \in [-1, 1]$), can be defined for the duty cycles of small vectors. The duty cycles of the small vectors $\mathbf{S0}_1$, $\mathbf{S0}_2$, $\mathbf{S1}_1$, and $\mathbf{S1}_2$ (Figure 5.3), are then given by

$$d_{S0,1} = \frac{1 + x_{S0}}{2} d_{s0}, \quad d_{S0,2} = \frac{1 - x_{S0}}{2} d_{s0} \tag{5.6}$$

$$d_{S1,1} = \frac{1 + x_{S1}}{2} d_{s1}, \quad d_{S1,2} = \frac{1 - x_{S1}}{2} d_{s1} \tag{5.7}$$

while i_S is given by

$$i_S = x_{S0} d_{S0} I_{NP}(\mathbf{S0}_1) + x_{S1} d_{S1} I_{NP}(\mathbf{S1}_1) \tag{5.8}$$

For each value of θ, i_S can reach a certain highest (maximum) value, $i_{S,hi}$, as a result of setting

$$x_{S0} = \text{sign}(I_{NP}(\mathbf{S0}_1)) \quad \text{and} \quad x_{S1} = \text{sign}(I_{NP}(\mathbf{S1}_1)) \tag{5.9}$$

Use of

$$x_{S0} = -\text{sign}(I_{NP}(\mathbf{S0}_1)) \quad \text{and} \quad x_{S1} = -\text{sign}(I_{NP}(\mathbf{S1}_1)) \tag{5.10}$$

results in i_S taking a respective lowest (minimum) value, $i_{S,lo}$. In both Equations 5.9 and 5.10, (x_{S0}, x_{S1}) becomes equal to $(-1, -1)$, $(1, -1)$, $(-1, 1)$, or $(1, 1)$, assigning the whole d_{S0} and d_{S1} to a certain small vector from the respective pair, according to Equations 5.6 and 5.7.

The locally averaged NP current, i_{NP}, is the sum of i_M and i_S:

$$i_{NP} = i_M + i_S \tag{5.11}$$

Given that i_M cannot be controlled by an NV strategy, the highest and lowest values, $i_{NP,hi}$ and $i_{NP,lo}$, of i_{NP} correspond to $i_{S,hi}$ and $i_{S,lo}$, respectively.

5.3.2 Effect of Switching Constraints

By adjusting the distribution factors of the small vectors, i_{NP} can take any value between $i_{NP,lo}$ and $i_{NP,hi}$. It can be shown, however, that freedom in this adjustment can lead to increased effective switching frequency. The values of x_{S0} and x_{S1} are therefore normally restricted using sets of switching constraints, which differ among NV strategies. For example, in the case of the NTV strategy, x_{S0} and x_{S1} can only take the values of ± 1. The switching sequences defined according to this restriction are shown in Table 5.1 [10]. The sequences are given for the first sextant, and $\mathbf{S0}_1$ and $\mathbf{S1}_1$ correspond to vectors 100 and 221, respectively (see Figures 5.2 and 5.3).

Table 5.1 Duty cycle distribution factors and switching sequences for the NTV strategy

Triangle	x_{S0}	x_{S1}	Switching sequence	Steps
tr_1	+1	n/a	100-200-210-200-100	4
	−1	n/a	200-210-211-210-200	4
tr_2	+1	+1	100-210-221-210-100	8
	+1	−1	100-110-210-110-100	4
	−1	+1	210-211-221-211-210	4
	−1	−1	110-210-211-210-110	4
tr_3	n/a	+1	210-220-221-220-210	4
	n/a	−1	110-210-220-210-110	4
tr_4	+1	+1	100-111-221-111-100	8
	+1	−1	100-110-111-110-100	4
	−1	+1	111-211-221-211-111	4
	−1	−1	110-111-211-111-110	4

Most sequences require four (instead of six) switching steps, therefore decreasing the effective switching frequency for the NTV strategy. In triangles tr_2 and tr_4, however, there are also sequences that incorporate eight switching steps. These are available, so that i_{NP} can always attain the values $i_{NP,lo}$ and $i_{NP,hi}$, by selecting x_{S0} and x_{S1} according to Equations 5.9 and 5.10.

Other NV strategies can forbid the eight-step sequences by imposing additional switching constraints in triangles tr_2 and tr_4. Then, $i_{NP,lo}$ and $i_{NP,hi}$ become unattainable for certain values of θ (which depend on ϕ and m). Given the set of switching constraints, SC, of such a strategy, then $i_{NP,hi|SC}$ and $i_{NP,lo|SC}$ can be defined as the highest and lowest values, respectively, of i_{NP}, that can be achieved while abiding by SC. It is noted here that the switching constraints of SC for triangles tr_1 and tr_3 are assumed to allow the values of ± 1 for x_{S0} and x_{S1}, as happens in the case of the NTV strategy. This is an essential requirement if the strategy that uses SC wishes to have a control over the NP voltage. NV strategies that do not fulfill it, such as the sinusoidal pulse width modulation (SPWM) or the SVM with equal duty cycle distribution among the two small vectors of each pair, have increased values of NP voltage ripple [6, 27].

Generally, the following inequality holds:

$$i_{NP,hi} \geq i_{NP,hi|SC} \geq i_{NP,lo|SC} \geq i_{NP,lo} \tag{5.12}$$

However, as the modulation index increases, $\mathbf{V}_{REF}$ spends a smaller fraction of the fundamental period in tr_2 and tr_4, where the additional switching constraints apply. For $m = 1$, this fraction is zero, thus $i_{NP,hi|SC}$ and $i_{NP,lo|SC}$ become equal to $i_{NP,hi}$ and $i_{NP,lo}$, respectively, during the entire cycle.

The analysis presented in the following sections assumes no switching constraints imposed on the duty cycle distribution factors, in order to derive the operational limits of the proposed concept. It can be adapted to a strategy that uses a given SC, if $i_{NP,hi}$ and $i_{NP,lo}$ are substituted by $i_{NP,hi|SC}$ and $i_{NP,lo|SC}$, respectively. It is important to note, however, that since $i_{NP,hi|SC}$ and $i_{NP,lo|SC}$ approximate $i_{NP,hi}$ and $i_{NP,lo}$ when m approaches 1, the presented results that refer to these values of m can be used irrespective of SC.

5.3.3   Zero-Ripple Region

The NP voltage in the NPC inverter is determined by the integral of the NP current. Thus, v_{NP} can be controlled to remain the same (at the end of each switching cycle) throughout the fundamental cycle, if i_{NP} can become equal to zero for all values of θ. Since i_{NP} is restricted between $i_{NP,lo}$ and $i_{NP,hi}$, this is possible when

$$(i_{NP,lo} \leq 0 \text{ and } i_{NP,hi} \geq 0) \tag{5.13}$$

holds for $\theta = [0, 360)°$. Examination of Equation 5.13 at different values of θ and m gives the zero-ripple region for NV strategies, first presented in [6] and shown later, in Figure 5.7.

5.3.4   A Lower Boundary for the NP Voltage Ripple

Figure 5.4 plots $i_{NP,lo}$, $i_{NP,hi}$, and i_M, according to Equations 5.5 and 5.8–5.11, during a fundamental cycle. A balanced set of sinusoidal output currents with rms value $I_o = 1$ A is assumed, which lag the phase voltages by 30° ($\varphi = -30°$), while m is set to 0.9. The waveforms of $i_{NP,lo}$ and $i_{NP,hi}$ cross the zero axis, therefore Equation 5.13 does not hold for certain values of θ, and NP voltage ripple is expected to appear. The existence of voltage ripple cannot be avoided by any NV strategy, but its peak value depends on the use of x_{S0} and x_{S1}.

While Equation 4.13 holds during the fundamental cycle, x_{S0} and x_{S1} can be adjusted to set i_{NP} to zero ($i_S = -i_M$). In this case, the small vectors fully compensate for the charge taken from the NP by the medium vectors, thus avoiding a change in the NP voltage. As soon as Equation 5.13 ceases to hold, at an angle θ_A, i_{NP} can no longer be set to zero, therefore charge starts to be taken from the NP. This carries on until an angle θ_B, when Equation 5.13 begins to hold again. Nevertheless, the charge ΔQ_{AB}, taken from the NP during $[\theta_A, \theta_B]$ can be minimized if i_{NP} is adjusted to its – in absolute terms – minimum value, which is $i_{NP,hi}$ for this interval:

$$\Delta Q_{AB,min} = \frac{1}{2\pi f} \int_{\theta_A}^{\theta_B} i_{NP,hi} d\theta \tag{5.14}$$

Since $i_M < 0$ during $[\theta_A, \theta_B]$, the above adjustment ($i_S = i_{S,hi}$) corresponds to the small vectors providing the maximum possible degree of compensation against the medium vectors, while $\Delta Q_{AB,min}$ corresponds to the minimum achievable uncompensated amount of charge. The integral term of Equation 4.14 is equal to the area of the shaded region between points A and B in Figure 5.4, if θ is expressed in radians. The absolute value of $\Delta Q_{AB,min}$ is inversely proportional to the reference frequency, while its sign is the same as the sign of i_M.

Since the two DC-link capacitors, as seen from the NP, are connected in parallel [10], $\Delta Q_{AB,min}$ causes an NP voltage variation, $\Delta V_{AB,min}$, given by Equation 5.15. The equation contains a negative sign, indicating that v_{NP} decreases (or increases) as a result of a positive (or negative) value of $\Delta Q_{AB,min}$:

$$\Delta V_{AB,min} = -\frac{\Delta Q_{AB,min}}{2C} \tag{5.15}$$

The key fact is that no NV strategy can prevent $\Delta Q_{AB,min}$ from leaving the NP. Thus, $\Delta V_{AB,min}$ is a lower limit for the NP voltage variation that will occur during the interval $[\theta_A, \theta_B]$. However, if $\Delta V_{AB,min}$ appears during an interval of the fundamental cycle, then the peak–peak NP voltage variation during the whole cycle cannot be lower than $|\Delta V_{AB,min}|$. Consequently, $|\Delta V_{AB,min}|$ is a lower boundary, $\Delta V_{NP,min}$, for the peak–peak NP voltage ripple that can be generated by an NV modulation strategy.

The same analysis can be applied to all intervals of the fundamental cycle in which Equation 5.13 does not hold (shaded in Figure 5.4). Those will be referred to as uncontrollable intervals (UIs), because during them the NP voltage is unavoidably driven by the uncompensated charge of the medium vectors. On the contrary, intervals in which Equation 5.13 does hold, will be referred to as controllable intervals (CIs), since during them the small vectors can be used to keep the NP voltage constant, or control it to some extent. If n uncontrollable intervals, UI_1, UI_2, $\ldots$, UI_n, appear during a fundamental cycle, having respective minimum NP voltage variations of $\Delta V_{UI1,min}$, $\Delta V_{UI2,min}$, $\ldots$, $\Delta V_{UIn,min}$, then

$$\Delta V_{NP,min} = \max\{|\Delta V_{UIk,min}|\}, k = 1\ldots n. \tag{5.16}$$

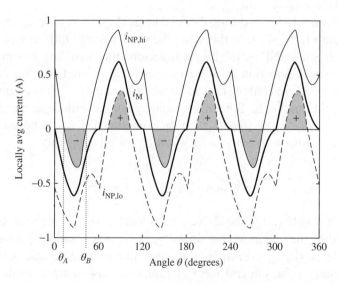

Figure 5.4 $i_{NP,lo}$, $i_{NP,hi}$, and i_M during a fundamental cycle, for $\phi = -30°$ and $m = 0.9$ ($I_o = 1$ A)

As dictated by Equations 5.14 and 5.15, $\Delta V_{NP,min}$ is a function of C and f, as well as I_o, ϕ, and m, since the waveforms of $i_{NP,lo}$ and $i_{NP,hi}$ change with these parameters. For given values of the above, an NV strategy that achieves $\Delta V_{NP,min}$ is optimal with respect to the NP voltage ripple. This lower boundary, on the other hand, may not be attainable in cases of successive UIs with same-sign NP voltage deviations, which will be discussed in Section 5.4.3.

5.4 Proposed Band-NV Strategies

This section investigates the effect of the criterion used in Step 3 (see Section 5.3) for minimizing the NP voltage ripple.

5.4.1 Criterion Used by Conventional NV Strategies

The two main objectives of an NPC converter modulation strategy in relation to the DC-link are the ability to bring the capacitors back to balance after transient imbalances and the minimization of the NP voltage ripple. The existing NV strategies try to achieve these objectives by decreasing any voltage imbalance that appears between the two DC-link capacitors. This is obtained by measuring the capacitor voltages and adjusting x_{S0} and x_{S1}, so that i_{NP} in Equation 5.11 will charge (or discharge) the capacitor having less (or more) voltage, respectively [10, 12, 21, 27]. Equivalently, the criterion conventionally used by the NV strategies for achieving NP balancing can be stated as follows:

Adjust i_{NP} to drive v_{NP} as close as possible to zero.

This criterion, which will be referred to as the conventional criterion, is suitable for bringing the capacitors back to balance after transient imbalances. Furthermore, it intuitively also seems to lead to the minimum NP voltage ripple. However, a simple example based on the analysis of the previous section can prove that the latter is not actually true.

Continuing the example of Figure 5.4, Figure 5.5a illustrates how the NP voltage, $v_{NP,Conv}$, varies during a third of a fundamental cycle when the conventional criterion is used. The NP voltage deviates from zero during the two UIs, shown in Figures 5.4 and 5.5b. The locally averaged NP current, $i_{NP,Conv}$, provides maximum compensation during these intervals, resulting in changing the NP voltage, of $\Delta V_{NP,min}$ and $-\Delta V_{NP,min}$, respectively. Moreover, $i_{NP,Conv}$ is adjusted according to the conventional criterion, to decrease $v_{NP,Conv}$ down to zero during the CIs. As a consequence, the changes of $\pm\Delta V_{NP,min}$ begin from $v_{NP,Conv} = 0$, giving a peak–peak value of $2\Delta V_{NP,min}$ to the NP voltage ripple.

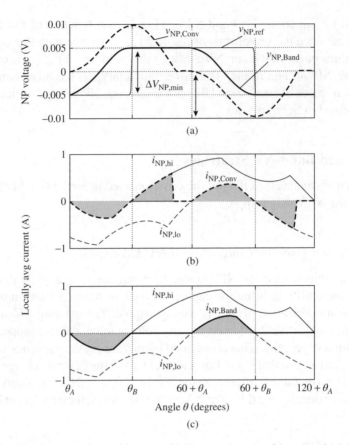

Figure 5.5 (a) $v_{NP,Conv}$, $v_{NP,Band}$, and $v_{NP,ref}$, (b) $i_{NP,Conv}$, $i_{NP,lo}$, and $i_{NP,hi}$, and (c) $i_{NP,Band}$, $i_{NP,lo}$, and $i_{NP,hi}$ during a third of a fundamental cycle, for $\phi = -30°$ and $m = 0.9$ ($I_o = 1$ A, $f = 1$ Hz, $C = 1$ F)

5.4.2   Proposed Criterion

The above increase in the value of NP voltage ripple can be avoided by a different criterion, which relies on the following observation: The voltage deviations have the same amplitude, but their sign changes successively from positive to negative and vice versa ($\Delta V_{UI1} = -\Delta V_{UI2} = \Delta V_{NP,min}$). Therefore, the NP voltage could periodically be driven to zero, even if i_{NP} was not used for this purpose. Instead, i_{NP} can be used as follows:

- **During UIs:** Provide the maximum possible charge compensation against the medium vectors. Namely, adjust x_{S0} and x_{S1} to achieve $i_{NP} = i_{NP,lo}$ if $i_M > 0$, or $i_{NP} = i_{NP,hi}$ if $i_M < 0$. The resulting NP voltage change will be $-\Delta V_{NP,min}$ and $\Delta V_{NP,min}$, respectively.
- **During CIs:** Adjust i_{NP} to (drive and) keep v_{NP} equal to $\Delta V_{NP,min}/2$, or $-\Delta V_{NP,min}/2$, if the i_M was negative, or positive, respectively, during the last UI.

The above can be achieved by periodically changing the NP reference voltage, $v_{\text{NP,ref}}$, as follows:

Adjust i_{NP} to drive v_{NP} as close as possible to $v_{\text{NP,ref}}$, which changes to $\Delta V_{\text{NP,min}}/2$, or $-\Delta V_{\text{NP,min}}/2$, at the end of an uncontrollable interval where i_{M} was negative, or positive, respectively.

This criterion will be referred to as the band criterion because it restricts the NP voltage, $v_{\text{NP,Band}}$, within a band that is centered around zero and has a width of $\Delta V_{\text{NP,min}}$. The waveforms of $v_{\text{NP,ref}}$ and $v_{\text{NP,Band}}$ are illustrated in Figure 5.5a. For this operating point, the peak–peak value of the NP voltage ripple is decreased to half. Nearest-vector strategies that operate according to the band criterion will be referred to as band-NV strategies.

5.4.3 Regions of Operation

The formulation of the band criterion relied on the observation that the change in NP voltage is opposite for every two successive UIs. As shown below, this is true for the greatest part of the NPC converter's operating range, but not for the entire range (defined by ϕ and m). The ϕ–m plane can be divided into regions, characterized by the number of UIs that appear per half cycle of i_{M}:

- **Region 0:** There are no UIs. $\Delta V_{\text{NP,min}}$ is equal to zero (zero-ripple region).
- **Region 1:** There is a single uncontrollable interval, UI_1, during each positive half cycle of i_{M}, with $\Delta V_{\text{UI1}} = -\Delta V_{\text{NP,min}} < 0$. The next uncontrollable interval, UI_2, appears in the negative half cycle of i_{M}, with $\Delta V_{\text{UI2}} = \Delta V_{\text{NP,min}} > 0$.
- **Region 2:** There are two uncontrollable intervals, UI_1 and UI_2 during each positive half cycle of i_{M}, with $\Delta V_{\text{UI1}} < 0$ and $\Delta V_{\text{UI2}} < 0$.

The example in Figure 5.4 belongs to Region 1. Figure 5.6a,b illustrate representative examples for Regions 0 and 2, respectively, while Figure 5.7 depicts the three Regions on the ϕ–m plane.

In Region 0, $v_{\text{NP,ref}}$ remains equal to zero (because $\Delta V_{\text{NP,min}}$ is 0) and the band criterion takes the form of the conventional criterion. Thus, in this region, a band-NV strategy operates like a conventional one. In Region 1, $v_{\text{NP,ref}}$ keeps $v_{\text{NP,Band}}$ to its positive or negative extreme ($\pm \Delta V_{\text{NP,min}}/2$) during CIs, knowing that during the following UIs the uncompensated charge will take it to the opposite extreme. This is not the case, however, in Region 2, where there are pairs of successive UIs with the same sign of uncompensated charge. The two intervals cause NP voltage deviations toward the same direction, and as a consequence a modified approach is required.

According to Equations 5.16, a lower boundary for the peak–peak NP voltage ripple in Region 2 is given by

$$\Delta V_{\text{NP,min}} = \max\{|\Delta V_{\text{UI1,min}}|, |\Delta V_{\text{UI2,min}}|\} \qquad (5.17)$$

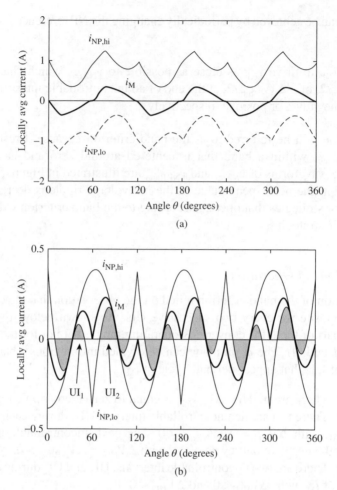

Figure 5.6 $i_{NP,lo}$, $i_{NP,hi}$, and i_M during a fundamental cycle, for (a) $\phi = -30°$, $m = 0.7$ (Region 0) and (b) $\phi = -3°$, $m = 1$ (Region 2)

To achieve this value, the effect of $\Delta V_{UI1,min}$ on the NP voltage should be partially (if $\Delta V_{UI1,min} > \Delta V_{UI2,min}$) or totally (if $\Delta V_{UI1,min} \leq \Delta V_{UI2,min}$) canceled during the controllable interval CI_{1-2}, found between UI_1 and UI_2. More precisely, during CI_{1-2}, $v_{NP,Band}$ should be driven to $\pm V_{CI1-2}$, where

$$V_{CI1-2} = \Delta V_{NP,min}/2 - |\Delta V_{UI2,min}| \qquad (5.18)$$

In this way, after the end of UI_2, $v_{NP,Band}$ will be equal to $\pm \Delta V_{NP,min}/2$, similar to what happens in Region 1. The following, modified version of the band criterion can be formed, to be used when the converter operates in Region 2:

Adjust i_{NP} to drive v_{NP} as close as possible to $v_{NP,ref}$, which changes to V_{CI1-2} or $-V_{CI1-2}$ at the end of UI_1, and to $\Delta V_{NP,min}/2$ or $-\Delta V_{NP,min}/2$ at the end of UI_2, if i_M during UI_1 and UI_2 was negative or positive, respectively.

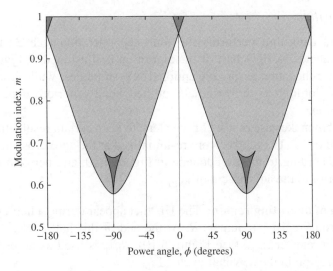

Figure 5.7 Regions of NPC converter operation: (white) Region 0, (gray) Region 1, and (black) Region 2

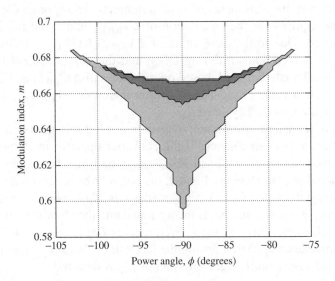

Figure 5.8 (White) Region 1, (gray) part of Region 2 where $\Delta V_{NP,min}$ can be achieved, and (black) part of Region 2 where $\Delta V_{NP,min}$ cannot be achieved

A simulation example illustrating the operation of the modified band criterion is presented in Section 5.6. The modified criterion can achieve an NP voltage ripple of $\Delta V_{NP,min}$ for the greatest part of Region 2, shown in Figure 5.8. In this part, i_{NP} can meet the implied requirement of being able to drive $v_{NP,Band}$ to $\pm V_{CI1\text{-}2}$ during $CI_{1\text{-}2}$ (before UI_2 begins). In the rest of Region 2, $\Delta V_{NP,min}$ cannot be attained by any NV strategy (see end of Section 5.3.4).

5.4.4   Algorithm

The proposed algorithm performs the four-step operation (see Section 5.3), for band-NV strategies. A flowchart of the algorithm is illustrated in Figure 5.9. Steps 1, 2, and 4 are the same as in conventional NV strategies, while in Step 3, $v_{NP,ref}$ is adjusted according to the original/modified band criterion, depending on the operating region.

The algorithm makes use of a set of registers, to keep certain results from the previous half cycle of i_M. The registers are re-initialized at the zero-crossings of i_M (once every $T/6$), according to the pseudocode in Figure 5.10, and perform the following operations, prior to the adjustment of $v_{NP,ref}$:

1. **Detection of operating region:** The UIs that appear during a half cycle of i_M are counted using register (counter) R. At the end of this half cycle, the value of R is stored in register R_{prev} (see Figure 5.10). R_{prev} is used as an estimate for the operating region in the new half cycle of i_M.
2. **Calculation of $V_{CI1\text{-}2}$ (for Region 2):** The algorithm makes use of the actual (minimized) NP voltage variation, $\Delta V_{NP}(1)$ and $\Delta V_{NP}(2)$ during the uncontrollable intervals UI_1 and UI_2, respectively, instead of $\Delta V_{NP,min}$. $\Delta V_{NP}(1)$ and $\Delta V_{NP}(2)$ are derived from measurements of v_{NP}, to indirectly incorporate the values of the parameters required for the calculation of $\Delta V_{NP,min}$ and avoid the integration of Equation 5.14. Due to this, the value of $\Delta V_{NP}(2)$ ($\Delta V_{UI2,min}$ in Equations 5.17 and 5.18) is not yet known during $CI_{1\text{-}2}$, and thus $V_{CI1\text{-}2}$ cannot be calculated. $V_{CI1\text{-}2,prev}$ is calculated instead, based on the values of $\Delta V_{NP}(1)$ and $\Delta V_{NP}(2)$ from the previous half cycle of i_M. The algorithm then uses $-V_{CI1\text{-}2,prev}$ as an estimate for $V_{CI1\text{-}2}$ in the new half cycle of i_M.
3. **Detection of transient imbalance:** The reference NP voltage should be set to zero in case of a transient imbalance. If the converter operates in Region 0, $v_{NP,ref}$ is already zero, and therefore a band-NV strategy performs balancing identically to a conventional one. In Regions 1 and 2, the NP voltage is expected to change sign during every half cycle of i_M. A register (flag), *Balflag*, which denotes that a sign change of v_{NP} was encountered, is incorporated into the algorithm. If *Balflag* is not found set following a sign change of i_M, register *Bal* is set to 0, indicating that a transient imbalance appeared during the last half cycle of i_M. The algorithm then keeps $v_{NP,ref}$ to zero, until a zero-crossing of v_{NP} is detected.

The values of R_{prev} and $V_{CI1\text{-}2,prev}$ are available at the beginning of each half cycle of i_M. Unless a transient imbalance has occurred, the algorithm continues with detecting the UIs using Equation 5.13. The NP voltage is sampled at the beginning and the end of each interval as $v_{NP,beg}$ and $v_{NP,end}$, respectively, and the NP voltage variation is stored in $\Delta V_{NP}(R) = v_{NP,end} - v_{NP,beg}$. At the end of the UIs, $v_{NP,ref}$ is updated as follows:

- If R_{prev} is 0, $v_{NP,ref}$ is set to zero by the initialization block. However, if an UI appears during the new half cycle of i_M, $v_{NP,ref}$ is set to $\Delta V_{NP}(1)/2$ at the end of it.

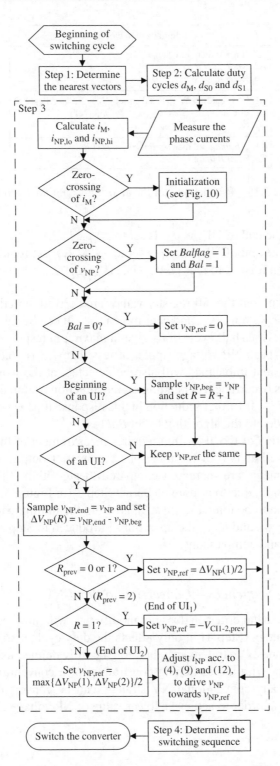

Figure 5.9   Flow chart of the switching loop of the proposed algorithm for the operation of band-NV strategies

Initialization block
A1. If $R = 0$ then set $v_{NP,ref} = 0$ A2. Set $R_{prev} = R$ A3. Reset R
B1. If $|\Delta V_{NP}(2)| > |\Delta V_{NP}(1)|$ then set $V_{CI1-2,prev} = -\Delta V_{NP}(2)/2$ else set $V_{CI1-2,prev} = \Delta V_{NP}(1)/2 - \Delta V_{NP}(2)$ B2. Reset $\Delta V_{NP}(1)$, $\Delta V_{NP}(2)$
C1. If $Balflag = 0$ then set $Bal = 0$ else set $Bal = 1$ C2. Reset $Balflag$

Figure 5.10   Pseudocode for the algorithm's initialization block

- If R_{prev} is 1, at the end of UI_1, $v_{NP,ref}$ is set to $\Delta V_{NP}(1)/2$.
- If R_{prev} is 2, at the end of UI_1, $v_{NP,ref}$ is set to $-V_{CI1-2,prev}$. It remains there until the end of UI_2, when it is set to $\max\{\Delta V_{NP}(1), \Delta V_{NP}(2)\}/2$.

It is worth pointing out that all register values are derived directly from the waveforms of i_M, $i_{NP,lo}$, $i_{NP,hi}$, and v_{NP}, and determine $v_{NP,ref}$. It can be shown that the use of results from the previous half cycle of i_M gives a maximum response time to a change in region or peak–peak NP voltage ripple, of less than $T/3$ ($= 6.67$ ms for a 50 Hz operation). A transient imbalance will also be detected, at the worst case, after $T/3$ from the time of its occurrence, depending on the direction of imbalance and its position in the cycle of i_M. If further reduction in this response time is needed, a threshold for $|v_{NP}|$ can be added to the algorithm to set Bal to 0.

The implementation of the algorithm requires measurement of the DC-link capacitor voltages, as well as two of the three-phase converter output currents. Its computational requirements are increased compared to algorithms that implement the conventional criterion, due to a more elaborate Step 3 (in Figure 5.9). However, this increase is expected to be tolerable since, apart from the multiplications required for the calculation of $i_{NP,lo}$ and $i_{NP,hi}$, the operations introduced in Step 3 are merely additions/subtractions and comparisons.

5.4.5   Switching Sequences – Conversion to Band-NV

The adoption of the band criterion does not modify the switching sequences followed by a given NV strategy (Step 4); it only affects the duty cycle distribution factors of small vectors. A band-NV strategy can therefore be fully defined using the switching sequences (together with the switching constraints) of a conventional NV strategy. Stated differently, any conventional NV strategy can be converted to operate as a band-NV strategy, in order to achieve a lower value of NP voltage ripple.

5.5   Performance of Band-NV Strategies

5.5.1   NP Voltage Ripple

The previous section showed how the adoption of a criterion, different from the DC-link capacitor imbalance (i.e., $v_{NP,ref} = 0$), for the duty cycle distribution of small vectors, creates the potential for decreasing the NP voltage ripple to its minimum possible value. In fact, it also proved that an NV strategy cannot obtain this minimum, unless the NP voltage at the beginning of the UIs is equal to its appropriate, positive or negative, extreme. The conventional NV strategies do not fulfill this constraint, as they constantly drive the NP voltage toward zero. Hence, a conventional NV strategy cannot obtain the minimum value of NP voltage ripple, in contrast to a band-NV strategy which potentially can.

Figure 5.11 plots the minimum amplitude of NP voltage ripple that can be achieved by conventional ($\Delta V_{NP,Conv}/2$) and band-NV strategies ($\Delta V_{NP,Band}/2$), as a function of ϕ and m. The presented values were derived similarly to [10] and are normalized according to

$$\frac{\Delta V_{NPn}}{2} = \frac{fC}{I_o} \frac{\Delta V_{NP}}{2} \tag{5.19}$$

The plots are shown only for $\phi = [-180, 0]°$, because they are repeated for $\phi = [0, 180]°$ ($\Delta V_{NP}(180 + \phi) = \Delta V_{NP}(\phi)$). Their values are the minimum achievable by each strategy type because they were derived assuming no switching constraints. Applying a set of switching constraints, SC, may have an effect on them. However, as shown in Section 5.3.2, when m approaches 1, $\Delta V_{NP|SC}$ approaches ΔV_{NP} for any imposed SC. Thus, $\Delta V_{NP,Conv}$ and $\Delta V_{NP,Band}$ for the critical value of $m = 1$ can provide the information required for DC-link capacitor sizing for any conventional and band-NV strategy, respectively. This argument is also supported by the equality of NP voltage ripple at $m = 1$, shown in [10] for the conventional NTV and symmetric strategies.

Moreover, the minimum NP voltage ripple achievable by band-NV strategies is equal to $\Delta V_{NP,min}$ in Regions 0 and 1, as well as in most of Region 2 (see Section 5.4.3). Thus, for these parts of the $\phi-m$ plane, Figure 5.11b is also a plot of $\Delta V_{NP,min}$, and can be re-derived using Equations 5.14 and 5.15. For the rest of the plane, $\Delta V_{NP,min}$ is unattainable by NV strategies.

In Figure 5.12 the values of Figure 5.11a,b are used to plot the percentage decrement, *decr*, of minimum NP voltage ripple in Regions 1 and 2 as:

$$decr = \left(1 - \frac{\Delta V_{NP,Band}}{\Delta V_{NP,Conv}}\right) \cdot 100\% \tag{5.20}$$

Based on this figure, it is worth identifying the range of loads in terms of ϕ, where the band-NV strategies can offer a remarkable decrement (*decr* $\geq 30\%$) of NP

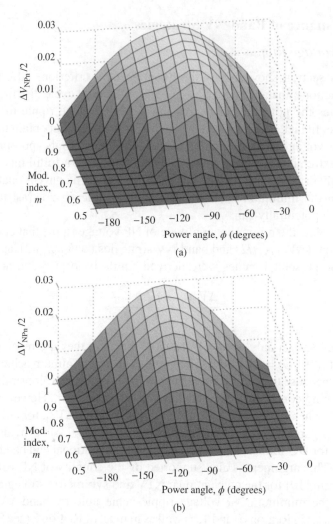

Figure 5.11 Minimum normalized amplitude of NP voltage ripple ($\Delta V_{NPn}/2$) for (a) conventional NV and (b) band-NV strategies

voltage ripple. Given that the converter should be able to operate at a high modulation index ($m = 0.95$), this range comprises the values of ϕ from approximately 0 to $-50°$ (and -140 to $-180°$). These values cover the use of the NPC converter as a grid connected inverter ($\phi \approx -30°$). On the contrary, the achievable decrement of NP voltage ripple for low power factor loads ($\phi \approx \pm 90°$) is smaller than 10%. Thus, band-NV strategies cannot offer a notable benefit in applications where the converter mainly has to provide reactive power.

Furthermore, it can be observed that *decr* is exactly equal to 50% for a large portion of Region 1. This is the region where the conventional NV strategies can hold the

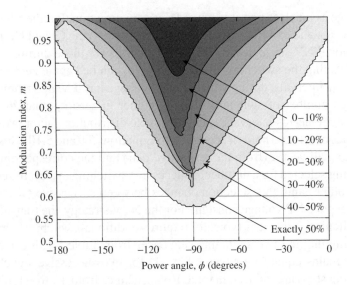

Figure 5.12   Value of *decr* as a function of ϕ and m

NP voltage to (approximately) zero for a part of the fundamental cycle. In this case, band-NV strategies can decrease the NP voltage ripple to half, as shown in the example of Figure 5.5.

5.5.2   *Effective Switching Frequency – Output Voltage Harmonic Distortion*

According to Equation 5.4, the converter's effective switching frequency is determined by the number of switching steps followed by the selected modulation strategy. The switching steps, in turn, are defined in the strategy's switching sequences. As explained in Section 5.4.5, each conventional NV strategy can be converted to a respective band-NV strategy having the same set of switching sequences. This conversion affects the duty cycle distribution of small vectors, but does not modify the switching sequences. Hence, the reduction in the NP voltage ripple offered by band-NV strategies, avoids the significant increase in the effective switching frequency caused by non-NV (or hybrid) strategies. Moreover, in contrast to the above strategies, band-NV strategies do not suffer from the output voltage harmonic distortion introduced by the use of non-NVs.

In comparison to the respective conventional strategies, on the other hand, there is an effect which increases the effective switching frequency of band-NV strategies when the converter operates in Region 1. The fundamental cycle can be divided into two types of interval, the duration of which is affected by the used criterion: (i) zero-i_{NP} intervals, during which i_{NP} should be kept to zero, so that the NP voltage remains constant and (ii) extreme-i_{NP} intervals during which i_{NP} should take the value $i_{NP,lo}$ or $i_{NP,hi}$, to drive the NP voltage as much as possible toward a certain direction. For the band-NV strategies, the zero-i_{NP} and extreme-i_{NP} intervals correspond to the CI and UI, respectively. For conventional NV strategies, the zero-i_{NP} intervals are those during which the NP voltage (neglecting the switching-frequency ripple) can be kept to zero, in contrast to the extreme-i_{NP} intervals, during which the NP voltage has deviated from zero.

The average number of switching steps is higher in zero-i_{NP} than in extreme-i_{NP} intervals. In extreme-i_{NP} intervals, each duty cycle distribution factor takes one of the extreme values of ± 1, and holds it according to Equations 5.9 and 5.10. This leads to the same, single small vector (from the respective pair) and the corresponding switching sequence being selected for a number of successive switching cycles. In zero-i_{NP} intervals, on the other hand, this selection changes between successive switching cycles, in order to keep the NP voltage constant. For the NTV strategy, the transition between different switching sequences typically requires additional switching steps [10]. For example, using the first two switching sequences from Table 5.1, it can be observed that, if x_{S0} remains equal to $+1$ (same for -1) for two successive switching cycles, eight switching steps are induced in total. If x_{S0} changes from $+1$ to -1, nine steps are induced, due to the transition from vector 100 to 200. If a strategy's switching constraints allow the use of intermediate values (between $+1$ and -1) for x_{S0} and x_{S1}, then during zero-i_{NP} intervals such a value is selected, leading to the use of both small vectors from the respective pair according to Equations 5.6 and 5.7. This again induces a higher number of switching steps compared to the extreme-i_{NP} intervals, where a single small vector is used.

The increased effective switching frequency for band-NV strategies arises from the fact that they have longer zero-i_{NP} intervals than the corresponding conventional strategies. This is because the band-NV strategies use the whole CIs as zero-i_{NP} intervals, whereas the conventional ones use parts of them as extreme-i_{NP} intervals, to drive the NP voltage toward zero. The degree of the above increase, as well as its effect on the converter design, will be discussed in Section 5.7.

5.6 Simulation of Band-NV Strategies

The previous sections showed that a decrement in NP voltage ripple can be achieved by the use of band-NV strategies for the NPC converter. The results did not refer to a specific band-NV strategy, since no set of switching sequences was assumed. As an example, this section compares the NTV modulation strategy with the respective band-NV strategy, referred to as band-NTV.

The two strategies were simulated using a MATLAB®-Simulink model of an NPC inverter with the following parameters: $V_{DC} = 1.8\,\text{kV}$, $C = 0.5\,\text{mF}$, $f_{sw} = 10\,\text{kHz}$, $f = 50\,\text{Hz}$. The load is varied between the presented simulations to attain the desired value of ϕ and an output rms current of 200 A. The current is kept to that value, so that the effect of m and ϕ on the amplitude of NP voltage ripple can be observed more clearly and verified against Figure 5.11, by means of Equation 5.19. The simulation figures illustrate the line−line voltage v_{ab}, current i_a, and capacitor C_1 reference voltage ($v_{C1,ref} = 900 + v_{NP,ref}$) when the inverter is modulated by the band-NTV strategy, as well as the voltage v_{C1} and the locally averaged currents $i_{NP,hi}$, $i_{NP,lo}$, and i_{NP}, for both strategies. See Appendix A for details of the simulation model.

The simulation results in Figures 5.13 and 5.14 assume a representative load power angle ϕ of $-30°$ (power factor of 0.866). The waveforms in Figure 5.13a,b correspond to two operating points, where *decr* is exactly equal to 50% and less than 50%, respectively. It can be noticed that, in the first case, the NTV strategy can drop the NP voltage down to zero during the CIs. As shown in Section 5.4, the band-NTV strategy can then halve the NP voltage ripple. In the second case, the decrement offered by the band-NTV strategy is less than 50%, but still significant (approximately 30%). The waveforms of $i_{NP,Conv}$ and $i_{NP,Band}$, which are responsible for the above decrements, can be studied based on Figure 5.5. Figure 5.13a, in particular, uses the same values

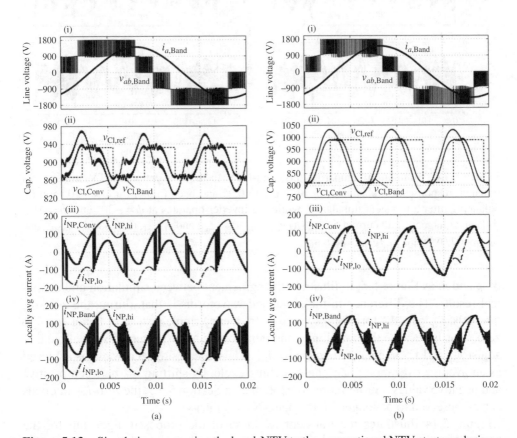

Figure 5.13 Simulation comparing the band-NTV to the conventional NTV strategy during a fundamental cycle, for $\phi = -30°$ and (a) $m = 0.9$ and (b) $m = 1$. (i) Line voltage v_{ab} and current $5 \times i_a$, (ii) $v_{C1,Conv}$, $v_{C1,Band}$, and $v_{C1,ref}$, (iii) $i_{NP,Conv}$, $i_{NP,lo}$, and $i_{NP,hi}$, and (iv) $i_{NP,Band}$, $i_{NP,lo}$, and $i_{NP,hi}$

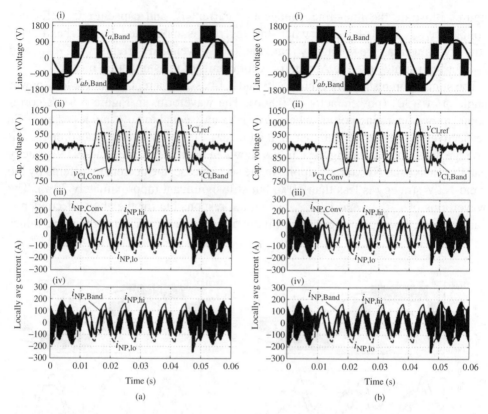

Figure 5.14 Simulation comparing the band-NTV to the conventional NTV strategy, for transient responses, (a) change of m from 0.7 to 0.95 and back to 0.7, for $\phi = -30°$ and (b) balancing after a transient imbalance, while the inverter operates at $\phi = -30°$ and $m = 0.95$. (i) Line voltage v_{ab} and current $5 \times i_a$, (ii) $v_{C1,Conv}$, $v_{C1,Band}$, and $v_{C1,ref}$, (iii) $i_{NP,Conv}$, $i_{NP,lo}$, and $i_{NP,hi}$, and (iv) $i_{NP,Band}$, $i_{NP,lo}$, and $i_{NP,hi}$

of ϕ and m as Figure 5.5, to illustrate the agreement between simulation and analytical results. The difference in the simulated waveforms is that, during the zero-i_{NP} intervals, i_{NP} gets the form of a switching-frequency ripple instead of being equal to zero. This is because the switching constraints of the NTV strategy do not allow it to adjust x_{S0} and x_{S1} to get a zero value for i_{NP}. Thus, the desired zero value is achieved as an average over more than one switching cycle, by shifting i_{NP} between positive and negative values. Furthermore, as explained in Section 5.5.2, the zero-i_{NP} intervals can be observed to be longer for the band-NTV strategy.

Figure 5.14 illustrates the transient response of the proposed algorithm for the band-NV strategies to two types of transient. In Figure 5.14a, the modulation index is changed from 0.7 to 0.95, and back to 0.7. The operating region changes, respectively, from Region 0 to 1 and back to 0, as determined by the waveforms of $i_{NP,lo}$ and $i_{NP,hi}$ (see Figures 5.5 and 5.6a). It can be noticed that it takes less than $T/3$ for the algorithm to detect each of the above changes and change $v_{NP,ref}$. Figure 5.14b presents a voltage balancing example after a forced transient imbalance. The imbalance is detected by the algorithm and is attenuated within the same time interval as in the case of the conventional strategy.

Figure 5.15 illustrates two simulation examples, where the inverter operates with a low and a high power-factor load, respectively. In the case presented in Figure 5.15a ($\phi = -83°$, $m = 0.95$), it can be observed that the band-NTV strategy does not decrease the NP voltage ripple significantly. This is because the inverter operates at the part of Region 1 where *decr* is less than 10% (see Figure 5.12). In Figure 5.15b ($\phi = -6°$, $m = 1$), the inverter operates in Region 2. The band-NTV strategy therefore uses the modified version of the band criterion, to re-adjust $v_{NP,ref}$ at the end of each of the two UIs. However, because $\Delta V_{NP}(2) > \Delta V_{NP}(1)$ in Figure 5.15b, $v_{NP,ref}$ does not change twice during each half cycle of i_M. Namely, $v_{NP,ref}$ takes the value of $\Delta V_{NP}(2)/2$ at the

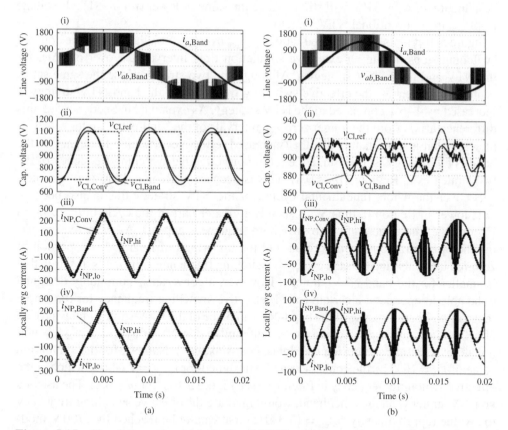

Figure 5.15   Simulation comparing the Band-NTV to the conventional NTV strategy during a fundamental cycle, for (a) $\phi = -83°$, $m = 0.95$ (Region 1) and (b) $\phi = -6°$, $m = 1$ (Region 2). (i) Line voltage v_{ab} and current $5 \times i_a$, (ii) $v_{C1,Conv}$, $v_{C1,Band}$, and $v_{C1,ref}$, (iii) $i_{NP,Conv}$, $i_{NP,lo}$, and $i_{NP,hi}$, and (iv) $i_{NP,Band}$, $i_{NP,lo}$, and $i_{NP,hi}$

end of each UI_2, and shifts to $-V_{CI1-2,prev}$, which is again equal to $\Delta V_{NP}(2)/2$, at the end of the following UI_1. The achieved decrement in NP voltage ripple for this case is 50%.

Figure 5.16a summarizes additional simulation results regarding the (normalized) amplitude of the NP voltage ripple generated by the NTV and band-NTV strategies, for the case of $\phi = -30°$. It also includes the cross-sections of Figure 5.11a,b for the above value of ϕ, which provide the minimum ripple achievable by conventional and band-NV strategies, respectively. The decrement in NP voltage ripple by the band-NTV strategy is 50% for up to $m = 0.925$ and gradually reaches a minimum of 31%. It is important to note that, according to Sections 5.3.2 and 5.5.1, this value of 31% is expected to be the same for all other NV strategies, as it corresponds to $m = 1$. For this case of $m = 1$, simulated in Figure 5.13b, the peak capacitor voltage is reduced from approximately 1040 to 995 V, which provides a significant advantage in terms of module voltage stress. The conversion to the band-NTV strategy can alternatively be used to reduce the DC-link capacitance, and thus the cost of the converter. A reduction of C by 31%, will still produce the same or lower (for $m < 1$) NP voltage ripple as the conventional NTV strategy.

Figure 5.16b plots the simulated ratio of $f_{sw,eff}$ over f_{sw} (i.e., the ratio of the counted switching steps during an extensive simulation time interval, Δt, over $6f_{sw}\Delta t$), for the NTV and band-NTV strategies. Although the two strategies use the same set of switching sequences, the effective switching frequency of the band-NTV strategy is increased by 4–6% compared to the NTV strategy. As explained in Section 5.5.2, this difference arises from the longer zero-i_{NP} intervals of the band-NTV strategy. Nevertheless, what is important is that the (increased) effective switching frequency of the band-NTV strategy is lower than the (common) effective switching frequency of the two strategies in Region 0 ($m \leq 0.8$). This is because in Region 0 the zero-i_{NP} intervals cover the whole fundamental cycle (for the NTV strategy this is also because m is lower in Region 0, thus $\mathbf{V}_{REF}$ spends a greater part of the cycle in tr_2 or tr_4 in Figure 5.3, where the eight-step switching sequences appear). As m increases, taking the operating point from Region 0 to the upper part of Region 1, the portion of zero-i_{NP} intervals, and thus the effective switching frequency, drops; it is this drop that is smaller for the band-NTV strategy. However, a converter is commonly designed to be able to operate at a certain (rated) effective switching frequency for the whole range of m. Therefore, use of the band-NTV in place of the NTV strategy may have an effect on the converter switching losses, but, unlike non-NV or hybrid strategies, will not affect its rating and design (modules, heat sink, etc.). Namely, for the simulation value of $f_{sw} = 10\,kHz$, the converter can be designed for $f_{sw,eff} = 7.6\,kHz$, which is the effective switching frequency in Region 0 (see Figure 5.16b, for $m \leq 0.8$). The use of a non-NV strategy, on the other hand, would increase the effective switching frequency to a value (approximately $f_{sw,eff} = 13.3\,kHz$) that cannot be reached by 1700 V modules which suit the assumed DC-link voltage level. According to Section 5.5.2, the above described effect is expected to appear in a similar way when converting other NV strategies to band-NV. Apart from the NTV, it has also already been verified for the case of the symmetric (NV) strategy [10].

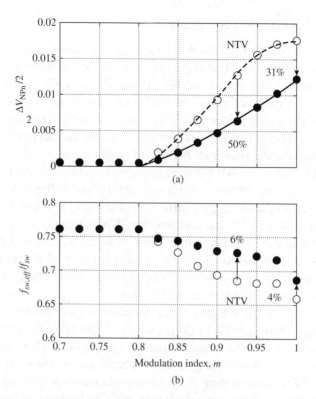

(a)

(b)

Figure 5.16 For $\phi = -30°$ and $m \geq 0.7$, (a) normalized amplitude of NP voltage ripple: (continuous line) $\Delta V_{NPn,Band}/2$, (dashed line) $\Delta V_{NPn,Conv}/2$, (filled circles) simulation for band-NTV strategy, (empty circles) simulation for NTV strategy and (b) ratio of $f_{sw,eff}$ over f_{sw}

Finally, the following two comments refer to aspects of the proposed concept that were not included in this study (which is also the case in similar studies [6, 10, 11, 17–21]):

- The switching-frequency NP voltage ripple was not considered during the formulation of band-NV strategies. However, at low switching frequencies relative to the fundamental, which are common for medium-voltage applications, this becomes important. In such cases, extra care should be taken when implementing a band-NV strategy, since the switching-frequency NP voltage ripple will affect the NP voltage sampling, and therefore the performance of the proposed algorithm.
- The calculation of duty cycles for the nearest vectors was based on [6], which assumes balanced capacitor voltages. However, when NP voltage ripple appears, the above duty cycles should be modified to avoid distortion (that is, injection of low-frequency harmonics) of the output voltage. Feed-forward techniques have been proposed for this purpose [8, 29, 30] which can be adapted for application on band-NV strategies.

5.7 Hybrid Modulation Strategies

The previous sections focused on NV strategies, proposing band-NV strategies to minimize the amplitude of NP voltage ripple in an NPC converter. As explained in Section 5.2, however, NV (and band-NV) strategies can only eliminate NP voltage ripple for a certain range of values of load power angle and converter modulation index. Non-NV strategies, on the other hand, can achieve NP voltage ripple elimination throughout the converter's operating range (i.e., for all values of ϕ and m), at the expense of increasing the switching losses and output voltage distortion.

The trade-off between the NP voltage ripple produced by NV strategies on the one hand, and the increment of switching losses and output voltage WTHD caused by non-NV strategies on the other, gave birth to hybrid strategies, which operate as combinations of the two. In [27], the well-known SPWM (which is an NV) strategy is combined with a non-NV strategy, implemented as a carrier-based PWM with two carrier waveforms. A variable, D, determines the fraction of the fundamental cycle where SPWM modulates the converter. It is important to note that, unless D is equal to zero, NP voltage ripple appears at the DC-link.

A different approach for creating a hybrid strategy can be found in [11]. There, the NTV strategy (named N3V in [11]) modulates the NPC converter in combination with a proposed non-NV strategy, named S3V. S3V is characterized by using three vectors during each switching cycle, of which, one is non-nearest. A threshold, $v_{NP,max}$, for the NP voltage is used to determine when S3V should be put into action ($|v_{NP}| > v_{NP,max}$), thus avoiding further NP voltage deviation (caused by the NTV). This approach also generates NP voltage ripple with amplitude $v_{NP,max}$. Moreover, if $v_{NP,max}$ is set close to zero with the aim of eliminating NP voltage ripple, successive transitions appear from the NV to the non-NV strategy and vice versa. As will be shown later (in Section 5.7.2), this effect is undesirable since it can increase the converter's switching losses.

Finally, in [31], a strategy that mitigates the drawbacks of the NTV^2 [14] is described. The strategy is classified here as hybrid, because the line voltage (PWM) waveforms in [14] indicate that it can partly operate as an NV strategy. Its formulation is based on a WTHD minimization process, which should be performed offline for nonlinear or imbalanced loads. For the case of linear and balanced loads, the results of the above process can be approximated by analytical equations. However, as explained in [16], an online estimator of the load power angle, as well as a detector for the linear and balanced nature of the load are still required.

This section proposes a straightforward way of creating hybrid strategies for the NPC converter, which have the following characteristics:

- Eliminate NP voltage ripple.
- Can operate with nonlinear or imbalanced loads.
- Can be built as combinations of any NV and non-NV strategy.

- Use the non-NV strategy to the minimum possible extent, thus minimizing the converter's switching losses and WTHD for the selected strategy combination.

5.7.1 Proposed Hybrid Strategies

The reader is reminded that in Section 5.3.4 the fundamental cycle of the NPC converter was divided into CI and UI, according to whether the following equation holds:

$$(i_{NP,lo} \leq 0 \text{ and } i_{NP,hi} \geq 0) \tag{5.21}$$

The above distinction can provide the basis for creating hybrid strategies which can eliminate NP voltage ripple. Given an NV strategy, X, and a non-NV strategy, Y, a hybrid strategy, H_{X-Y}, that combines the two can be built according to the flowchart in Figure 5.17. Namely, the converter can be modulated using X throughout the CIs, since during them, X is capable of holding v_{NP} to zero. During the UIs, on the other hand, Y should be put into action.

CIs and UIs can be identified in practice using the instantaneous (sampled) values of the phase currents as values of $I_{NP}(\mathbf{V})$ in Equations 5.5 and 5.8–5.10. In this way, operation of hybrid strategies according to Figure 5.17 is equally achievable for converter loads that draw non-sinusoidal or imbalanced currents. Moreover, it can be performed for loads that require non-sinusoidal voltages, since the NVs and duty cycles d_M, d_{S0}, and d_{S1}, can be determined for any, circular or not, movement of $\mathbf{V}_{REF}$ in the linear modulation region (circle in Figure 2.13). An example of such a load is a grid with a non-sinusoidal voltage, which the converter has to provide with sinusoidal currents.

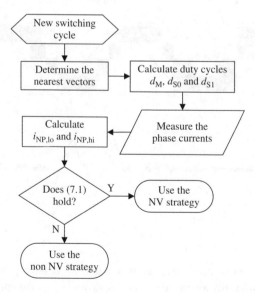

Figure 5.17   Flowchart for the proposed hybrid strategies

5.7.2 Simulation Results

An NPC inverter with $V_{DC} = 1.8\,\text{kV}$, $C_1 = C_2 = 0.5\,\text{mF}$, $f_{sw} = 8\,\text{kHz}$, and $f = 50\,\text{Hz}$ was simulated using MATLAB®-Simulink (SimPowerSystems Toolbox). The simulation figures illustrate the locally averaged currents $i_{NP,lo}$ and $i_{NP,hi}$, the applied modulation strategy, the line−line voltage v_{ab} and phase current(s), and the capacitor voltage v_{C1}. In Figure 5.18, the inverter supplies a linear and balanced load with $\phi = -30°$ (power factor of 0.866), while m is set to 0.9. The waveforms of $i_{NP,lo}$ and $i_{NP,hi}$ are shown for the entire simulation, but they are only used by the simulated hybrid strategy, according to Figure 5.17. This strategy, $H_{NTV-S3V}$, combines the (NV) NTV strategy [10], with the (non-NV) S3V strategy, proposed in [11]. In order to demonstrate its operation as compared to the combined strategies, the inverter is modulated for one

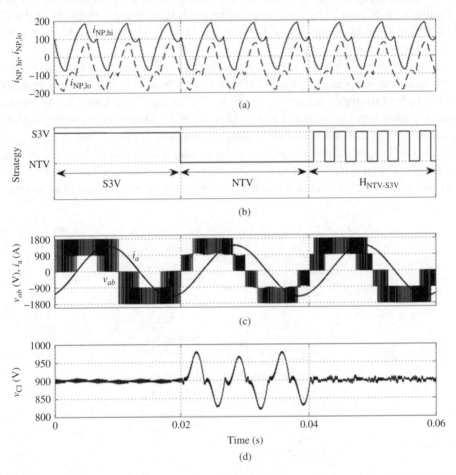

Figure 5.18 Simulation of NPC inverter modulated successively by the S3V, NTV, and $H_{NTV-S3V}$ strategies (a) locally averaged currents $i_{NP,lo}$ and $i_{NP,hi}$, (b) applied modulation strategy, (c) line voltage v_{ab} and current $5 \times i_a$, and (d) capacitor voltage v_{C1}

fundamental period (0.02 s) by each strategy, as follows: From 0 to 0.02 s by the S3V, from 0.02 to 0.04 s by the NTV, and from 0.04 to 0.06 s by the $H_{NTV-S3V}$ strategy. It can be observed that the low-frequency NP voltage ripple that appears when the NTV strategy modulates the inverter is eliminated by the $H_{NTV-S3V}$. This happens even though the NTV is still applied in place of the S3V for a significant part (48.5%) of the fundamental cycle. Moreover, the line voltage waveform generated by the $H_{NTV-S3V}$ can be seen to be enhanced (i.e., closer to the five-level waveform generated by the NTV strategy) as compared to the S3V strategy, thus having a decreased value of WTHD.

Combining the NTV with the S3V according to [11] and using a voltage threshold $v_{NP,max}$ of 5 V, has the effect shown in Figure 5.19. Due to the NTV strategy, the NP voltage quickly reaches the threshold and varies around it, thus causing multiple transitions between the two strategies. Such transitions, however, can introduce additional switching steps to the converter. For example, for $\mathbf{V}_{REF}$ as in Figure 5.2, the NTV strategy may need to use the switching sequence "210-110-100-110-210" (from Table 5.1). The S3V, on the other hand, would avoid the (ripple-generating) medium vector 210 and use "100-200-220-200-100" for a similar $\mathbf{V}_{REF}$. As a consequence, for each transition between the two strategies in that area of $\mathbf{V}_{REF}$, two additional switching steps will be introduced to the converter (to switch from 100 to 210 or vice versa).

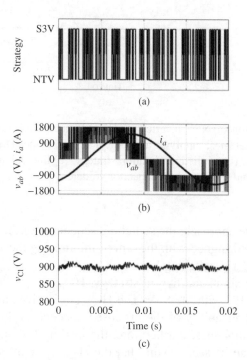

(a)

(b)

(c)

Figure 5.19 Simulation of NPC inverter modulated by a hybrid strategy combining the NTV and S3V. According to [11] ($v_{NP,max} = 5$ V) (a) applied modulation strategy, (b) line voltage v_{ab} and current $5 \times i_a$, and (c) capacitor voltage v_{C1}

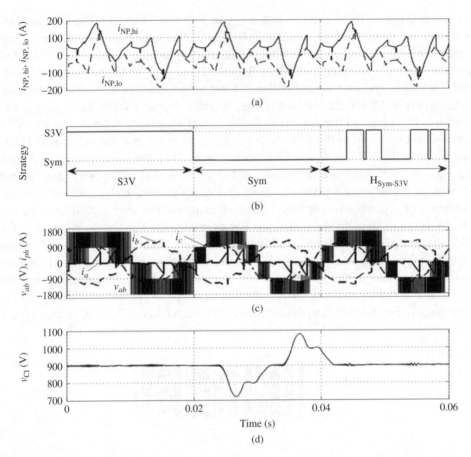

Figure 5.20 Simulation of NPC inverter modulated successively by the S3V, symmetric (Sym), and $H_{Sym-S3V}$ strategies, supplying a nonlinear and imbalanced load (a) locally averaged currents $i_{NP,lo}$ and $i_{NP,hi}$, (b) applied modulation strategy, (c) line voltage v_{ab} and three-phase currents $5 \times i_a$, $5 \times i_b$, and $5 \times i_c$, and (d) capacitor voltage v_{C1}

When using the above approach, the switching sequences of the combined strategies should therefore be redesigned with the aim of minimizing the added steps and their impact on the converter's switching losses.

In Figure 5.20, the simulated hybrid strategy, $H_{Sym-S3V}$, combines the symmetric (Sym) NV strategy [10] with the S3V. The $H_{Sym-S3V}$ is given as a second example, demonstrating the applicability of the proposed concept on different combinations of NV and non-NV strategies. Furthermore, the load in this simulation is nonlinear and imbalanced. It can be seen, again, that the $H_{Sym-S3V}$ eliminates the NP voltage ripple generated by the symmetric strategy, even though the latter is still applied for approximately 50% of the fundamental cycle.

Figure 5.21 plots the percentage duration of UIs during a fundamental cycle, as a function of ϕ and m. The presented values are derived using Equations 5.5 and 5.8–5.11,

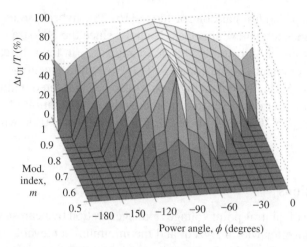

Figure 5.21 Percentage duration of uncontrollable intervals as a function of ϕ and m according to Equations 5.5 and 5.8–5.11, for sinusoidal and balanced phase currents

for the case of sinusoidal and balanced phase currents (they are shown for $\phi = -180°$ to $0°$; they are identical for $\phi = 0°$ to $180°$, respectively). For hybrid strategies created according to Figure 5.17, Figure 5.21 also depicts the percentage duration of applying the (selected) non-NV strategy. It can be observed that, for $\phi = -90°$ and $m > 0.7$, this percentage approaches 100%. Thus, for purely reactive loads, the proposed hybrid strategies offer no benefit compared to non-NV strategies, since they operate as such (the same has been observed in [31]). On the other hand, for less reactive loads, the participation of non-NV strategies is lower than 100% and decreases with m. For low values of m it drops to zero, therefore the hybrid strategies operate as NV strategies.

The proposed approach inherently guarantees minimum participation of non-NV strategies, since non-NV strategies are only applied when an NP voltage deviation cannot be prevented by NV strategies. Hence, for a given combination of strategies, the approach yields an NP-voltage-ripple-eliminating strategy with minimum switching losses and output voltage WTHD.

An estimate for these switching losses can be obtained by an analysis similar to that included in [27]. Such an analysis can be performed because the intervals of the fundamental cycle where each strategy is applied can be specified analytically (as in Figure 5.4); this is not the case in the approaches of [11, 31]. In comparison to [11], the proposed approach also has the advantage of avoiding multiple transitions between the combined strategies, and thus the need for modification of their switching sequences. In fact, operation according to Figure 5.17 can be readily added to any implemented pair of strategies. Computationally, this incurs little additional cost, since the determination of the NVs and the calculation of duty cycles (in Figure 5.17) are already in place for any NV strategy.

A last comment refers to the capacitor (or NP) balancing capabilities of the proposed strategies, after possible NP voltage deviations. Unlike non-NV strategies which may need to implement additional control loops [16, 28], hybrid strategies can rely on their NV strategies to achieve this task. Namely, since the NV strategies are applied during the CIs, they are given the chance to adjust i_{NP} in favor of capacitor balancing (a similar comment can be found in [11]). As the percentage duration of CIs increases (Figure 5.21), the balancing process can be completed within a shorter time interval. A simulation detail illustrating the balancing operation of $H_{Sym-S3V}$ can be observed in Figure 5.20 (see v_{C1} being driven to 900 V at 0.04 s, when the $H_{Sym-S3V}$ is applied).

5.8   Conclusions

The mechanism of neutral-point voltage ripple generation by nearest-vector strategies for the NPC converter was analyzed, and the minimum achievable ripple amplitude was derived. It was shown that conventional (that is, existing) NV strategies cannot attain this minimum, due to the criterion they use to perform NP (DC-link capacitor) balancing. A new criterion and a respective algorithm were proposed to form band-NV strategies, which provide this possibility.

Conventional NV strategies can also be converted to operate according to the proposed approach. Analytical results showed that the conversion of a conventional to a band-NV strategy can decrease the NP voltage ripple by approximately 30–50% ($\phi = -30°$). Alternatively, if a certain amplitude of NP ripple can be afforded, the DC-link capacitance (required to limit the NP voltage ripple) can be reduced by 30%. For higher power factor loads this percentage can reach closer to 50%, whereas no significant benefit can be expected for very low power factor loads (ϕ close to $\pm 90°$). The above results were verified by simulations of the well-known NTV strategy and its band-NV equivalent. It was also shown that similar outcomes are equally attainable by the conversion to band-NV of other conventional strategies.

Compared to non-NV and hybrid strategies, band-NV strategies offer the advantage of not affecting the converter's rated switching frequency. However, in comparison to the respective conventional NV strategies, a small increase in switching losses can be expected when the converter operates at a high modulation index.

This chapter also presented a simple concept for creating hybrid modulation strategies that can eliminate NP voltage ripple from the NPC converter. The proposed hybrid strategies can be based on existing NV and non-NV strategies, combining the two in a way that minimizes the converter's switching losses and output voltage harmonic distortion. The benefit offered in comparison to non-NV strategies (which also eliminate NP voltage ripple) increases with the load power factor and decreases with the converter modulation index. The performance of the proposed concept is also enhanced as compared to other approaches for the creation of hybrid strategies. Its operation was demonstrated for different combinations of strategies and with nonlinear and imbalanced loads.

References

1. Nabae, A., Takahashi, I. and Akagi, H. (1981) A new neutral-point-clamped PWM inverter. *IEEE Transactions on Industry Applications*, **IA-17** (5), 518–523.
2. Rodriguez, J., Bernet, S., Steimer, P.K. and Lizama, I.E. (2010) A survey on neutral-point-clamped inverters. *IEEE Transactions on Industrial Electronics.*, **57** (7), 2219–2230.
3. Teichmann, R. and Bernet, S. (2005) A comparison of three-level converters versus two-level converters for low-voltage drives, traction and utility applications. *IEEE Transactions on Industry Applications*, **41** (3), 855–865.
4. SEMIKRON (2009) Technical Information IGBT Modules, Datasheet SK100MLI066T, SEMIKRON, Nuremberg.
5. Infineon Technologies AG (2011) Technical Information IGBT Modules, Datasheet F3L300R07PE4, Infineon Technologies AG, Warstein.
6. Celanovic, N. and Boroyevich, D. (2000) A comprehensive study of neutral-point voltage balancing problem in three-level neutral-point-clamped voltage source PWM inverters. *IEEE Transactions on Power Electronics*, **15** (2), 242–249.
7. Kaminski, N. and Kopta, A. (2011) Failure Rates of HiPak Modules Due to Cosmic Rays, ABB Switzerland Ltd, Lenzburg. Appl. Note 5SYA2042-04.
8. Celanovic, N., Celanovic, I., and Boroyevich, D. (2001) The feedforward method of controlling three-level diode clamped converters with small dc-link capacitors. Proceedings of the 32nd Annual IEEE PESC, June 17–21, Vol. 3, pp. 1357–1362.
9. Celanovic, N. and Boroyevich, D. (2001) A fast space-vector modulation algorithm for multilevel three-phase converters. *IEEE Transactions on Industry Applications*, **37** (2), 637–641.
10. Pou, J., Pindado, R., Boroyevich, D. and Rodriguez, P. (2005) Evaluation of the low-frequency neutral-point voltage oscillations in the three-level inverter. *IEEE Transactions on Industrial Electronics*, **52** (6), 1582–1588.
11. Gupta, K. and Khambadkone, M. (2007) A simple Space Vector PWM scheme to operate a three-level NPC inverter at high modulation index including overmodulation region, with neutral point balancing. *IEEE Transactions on Industry Applications*, **43** (3), 751–760.
12. Zhang, H.B., Finney, S.J., Massoud, A. and Williams, B.W. (2008) An SVM algorithm to balance the capacitor voltages of the three-level NPC active power filter. *IEEE Transactions on Power Electronics*, **23** (6), 2694–2702.
13. Lewicki, A., Krzeminski, Z. and Abu-Rub, H. (2011) Space-vector pulse width modulation for three-level NPC converter with the neutral point voltage control. *IEEE Transactions on Industrial Electronics*, **58** (11), 5076–5086.
14. Busquets-Monge, S., Bordonau, J., Boroyevich, D. and Somavilla, S. (2004) The nearest three virtual space vector PWM-A modulation for the comprehensive neutral-point balancing in the three-level NPC inverter. *IEEE Power Electronic Letters*, **2** (1), 11–15.
15. Pou, J., Zaragoza, J., Rodríguez, P. *et al.* (2007) Fast-processing modulation strategy for the neutral-point-clamped converter with total elimination of the low-frequency voltage oscillations in the neutral point. *IEEE Transactions on Industrial Electronics*, **54** (4), 2288–2299.
16. Busquets-Monge, S., Ortega, J.D., Bordonau, J. *et al.* (2008) Closed-loop control of a three-phase neutral-point-clamped inverter using an optimized virtual-vector-based pulsewidth modulation. *IEEE Transactions on Industrial Electronics*, **55** (5), 2061–2071.

17. Newton, C. and Sumner, M. (1997) Neutral point control for multi-level inverters: theory, design and operational limitations. Proceedings of the IEEE Industry Applications Society Annual Meeting, New Orleans, LA, pp. 1336–1343.

18. Ogasawara, S. and Akagi, H. (1993) Analysis of variation of neutral point potential in neutral-point-clamped voltage source PWM inverters. Conference Records of IEEE IAS Annual Meeting, pp. 965–970.

19. Bruckner, T. and Holmes, D.G. (2005) Optimal pulse-width modulation for three-level inverters. *IEEE Transactions on Power Electronics*, **20** (1), 82–89.

20. Wang, C. and Li, Y. (2010) Analysis and calculation of zero-sequence voltage considering neutral-point potential balancing in three-level NPC converters. *IEEE Transactions on Industrial Electronics*, **57** (7), 2262–2271.

21. Pou, J., Zaragoza, J., Ceballos, S. *et al.* (2012) A carrier-based PWM strategy with zero-sequence voltage injection for a three-level neutral-point-clamped converter. *IEEE Transactions on Power Electronics*, **27** (2), 642–651.

22. Tallam, R.M., Naik, R. and Nondahl, T.A. (2005) A Carrier-Based PWM scheme for neutral-point voltage balancing in three-level inverters. *IEEE Transactions on Industry Applications*, **41** (6), 1734–1743.

23. Videt, A., Le Moigne, P., Idir, N. *et al.* (2007) A new carrier-based PWM providing common-mode-current reduction and dc-bus balancing for three-level inverters. *IEEE Transactions on Industrial Electronics*, **54** (6), 3001–3011.

24. Wang, F. (2002) Sine-triangle versus space-vector modulation for three-level PWM voltage-source inverters. *IEEE Transactions on Industry Applications*, **38** (2), 500–506.

25. Da-peng, C., Wen-xiang, S., Hui, X.I. *et al.* (2009) Research on zero-sequence signal of space vector modulation for three-level neutral-point-clamped inverter based on vector diagram partition. IEEE 6th International Power Electronics Motion Control Conference, Vol. 3, pp. 1435–1439.

26. Rodriguez, J., Franquelo, L.G., Kouro, S. *et al.* (2009) Multilevel converters: an enabling technology for high-power applications. *Proceedings of the IEEE*, **97** (11), 1786–1817.

27. Zaragoza, J., Pou, J., Ceballos, S. *et al.* (2009) A comprehensive study of a hybrid modulation technique for the neutral-point-clamped converter. *IEEE Transactions on Industrial Electronics*, **56** (2), 294–304.

28. Zaragoza, J., Pou, J., Ceballos, S. *et al.* (2009) Voltage-balance compensator for Carrier-Based Modulation in the neutral-point-clamped converter. *IEEE Transactions on Industrial Electronics*, **56** (2), 305–314.

29. Pou, J., Boroyevich, D. and Pindado, R. (2002) New feedforward space-vector PWM method to obtain balanced AC output voltages in a three-level neutral-point-clamped converter. *IEEE Transactions on Industrial Electronics*, **49** (5), 1026–1034.

30. Leon, J.I., Vazquez, S., Portillo, R. *et al.* (2009) Three-dimensional feedforward space vector modulation applied to multilevel diode-clamped converters. *IEEE Transactions on Industrial Electronics*, **56** (1), 101–109.

31. Busquets-Monge, S., Somavilla, S., Bordonau, J. and Boroyevich, D. (2007) Capacitor voltage balance for the neutral-point-clamped converter using the virtual space vector concept with optimized spectral performance. *IEEE Transactions on Power Electronics*, **22** (4), 1128–1135.

6

Digital Control of a Three-Phase Two-Level Grid-Connected Inverter

6.1 Introduction

Power electronic inverters are commonly used to connect small distributed generation systems (e.g., micro-CHP (combined heat and power), PV (photovoltaic), and small wind turbines) to the grid [1–7], as discussed in Chapter 1. These inverters are required by standards and engineering recommendations [8–12] to control the quality of the current fed into the grid but must not regulate the voltage at the common point of coupling.

In recent years there has been significant research into the control of grid-connected inverters [4–7, 13–17]. Various filter topologies and controller structures have been proposed and analyzed. The LCL filter topology is commonly used for its advantage over series inductor filters [6] in terms of lower switching frequency and/or smaller inductor size, but it has the disadvantage of presenting a lower impedance to utility harmonics, in addition to requiring a more complex control scheme to deal with filter resonance. For simplicity and guaranteed stability, many authors [5, 7, 13, 16, 18, 19] chose to control the inverter current before the filter rather than the grid current. Such systems can, however, suffer from problems arising from filter resonance, such as underdamped transient response oscillations, large overshoot, and oscillations induced by utility harmonics near the resonance frequency. Reference [18] proposed a lead-lag compensating loop of the filter capacitor voltage to actively damp filter resonance. Resistors connected in series with capacitors or inductors are also sometimes used, although this is not an efficient option. Direct feedback of the actual grid current can be shown to result in an unstable system, and it is necessary to add an inner feedback loop, usually of capacitor current, to ensure system stability [4, 14, 20].

Although many authors reported that they implemented their grid-connected inverter system controllers using microprocessors or DSPs (digital signal processors), there is usually little discussion of the details of the practical digital implementation, regarding

Power Electronic Converters for Microgrids, First Edition. Suleiman M. Sharkh, Mohammad A. Abusara, Georgios I. Orfanoudakis and Babar Hussain.
© 2014 John Wiley & Sons, Ltd. Published 2014 by John Wiley & Sons, Ltd.
Companion Website: www.wiley.com/go/sharkh

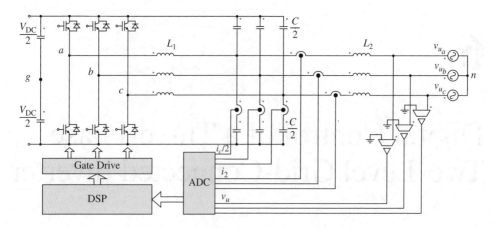

Figure 6.1   Grid-connected inverter system

the sampling strategy or the effect of time lag (computation time, A/D conversion, sampling delay time), inherent in discrete time systems, on the system stability and quality of the output current.

Figure 6.1 shows a circuit diagram of the inverter system with output LCL filter. The operating parameters and component values of the system are shown in Table 6.1. There are different possible options for the output filter structure. A filter with only an L_1 inductor has the advantage of being simple with fewer components. However, the inductor needs to be relatively large to limit the ripple current; consequently, the system dynamics will be poor due to the large voltage drop across the inductor. A smaller value of L_1 can be used in an LC filter structure, and by increasing the capacitance C, the inductance L_1 can be further reduced for the same attenuation. The presence of a capacitor will, however, create a resonance problem that needs to be dealt with, either passively by using a resistor, or actively by using an extra feedback loop of the capacitor current, as will be shown in this chapter. A very high capacitance value is not recommended as it reduces the output impedance of the filter, thus providing an easy path for harmonic currents caused by grid voltage harmonics. In an LCL filter structure, the second inductor L_2 increases the impedance seen by the grid, especially in places where the grid impedance is very low, and hence prevents the ripple current from being injected into the grid. Also L_2 limits the capacitor inrush charging current during starting.

The selection of the filter component values is a trade-off between inductor size, conduction and switching losses in the IGBTs, resonance frequency, and the degree of attenuation of the switching frequency ripple. The inductor size depends mainly on the value of the current, the switching frequency, and the inductance value. For the given power rating in Table 6.1, the switching frequency is limited by the maximum IGBTs power dissipation to 10 kHz. The L_1 inductance of 300 μH has been chosen to limit the ripple current to a value of ~ 20% of the rated current, as dictated by the ratings

Table 6.1 System parameter and component values

Rated power	100 kV A
Utility phase voltage	230 V (rms)
Grid frequency	50 Hz
DC-link voltage	750 V (dc)
Switching frequency	10 kHz
Sampling frequency	20 kHz
Inductor L_1	300 µH
Inductor L_2	20 µH
Capacitor C	200 µF ($2 \times 100 \mu F$)
Time delay T_d	50 µs
Output current	140 A (rms)
Output current phase	0.0°

and losses in the IGBT modules. L_2 and C have to be chosen to satisfy two criteria: (i) The impedance of L_2 at the switching frequency needs to be several times bigger than the impedance of C in order to prevent the ripple current from being injected into the grid. (ii) The natural damped frequency ω_d of the system needs to be within the controlled range of the digital controller. As a rule of thumb, the sampling frequency (20 kHz) needs to be between 8 and 10 times the natural damped frequency of the system [21]. The natural damped frequency is given by $\omega_d = \omega_n \sqrt{1 - \zeta^2}$ where ω_n is the undamped natural frequency of the filter, and ζ is the damping ratio which is a characteristic of the controller and the filter and typically has a value between 0.3 and 0.7. For an LCL filter, the undamped natural frequency is given by $\omega_n = \sqrt{(L_1 + L_2)/L_1 L_2 C}$. C and L_2 have been chosen to be 200 µF and 20 µH, respectively, which results in $\omega_n = 16330$ rad s^{-1}. For a worst case scenario of $\zeta = 0.3$, $\omega_d = 15578$ rad s^{-1} which is nearly 1/8 of the switching frequency. Also, the impedance of L_2 at the switching frequency is nearly 16 times bigger than the impedance of C.

The proposed controller includes a grid current feedback loop, an inner capacitor current feedback loop to provide active damping, and a grid voltage feedforward loop to cancel out the grid voltage disturbance effect. The digital controller computational time delay is shown to reduce the stability of the inner loop. It is also shown that it is better, from the point of view of output current THD (total harmonic distortion) quality, to increase the inner loop time delay to a full one sampling period, so that the envelope mains frequency (nearly ripple free) component of the current is sampled at the peaks and troughs of the PWM (pulse width modulation), away from the switching instances (and associated noise) of the transistors. The effect of this time delay is mitigated by using an equivalent of a digital Smith controller; the inner loop current at the time of the PWM modulating signal update is predicted based on delayed measured values. The selection of the loop gains and outer loop compensation is discussed, and it is shown that the gains need to be selected to provide a compromise between transient stability and grid harmonic disturbance rejection. Simulation and experimental performance results are presented to validate the design.

6.2   Control Strategy

Figure 6.2a shows an equivalent linearized per phase circuit of the inverter. The voltage source v_{gn} is the voltage difference between the neutral point and the middle of the DC-link, which can be shown to be virtually zero if the filter capacitors are connected to the DC-link as shown in Figure 6.1 [20].

Figure 6.2b shows the per-phase controller and system block diagram assuming $v_{gn} = 0$. The controller aims to modulate the inverter voltage V_{mod} to regulate the magnitude and phase angle of the grid current so as to control the real and reactive power flow, while maintaining good current quality despite system nonlinearity, imbalance, and utility voltage harmonic disturbances. It can be readily shown that feedback of the grid current alone is unstable [14, 16] and it is necessary to add a minor feedback loop of the capacitor current to actively dampen filter resonance (this is equivalent to inserting a virtual resistance in series with the capacitor). The inner loop needs to have a high bandwidth to provide effective damping of filter resonance, and the stability of this loop is therefore sensitive to digital sampling time delay. To improve the stability

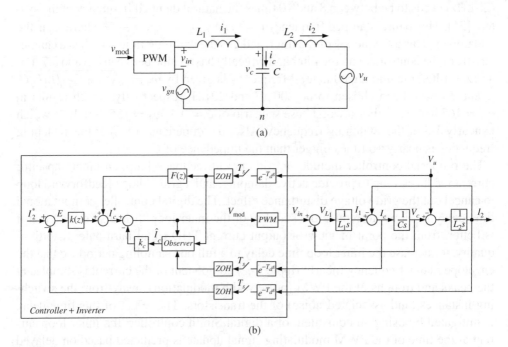

Figure 6.2   (a) Per phase equivalent circuit and (b) per phase block diagram of system and controller

of the inner loop, a capacitor current observer, whose design will be discussed later, is added to compensate for the time delay by predicting the value of the capacitor current at the moment of updating the modulating voltage V_{mod}. Sampling time delay, however, has a negligible effect on the stability of the outer loop, which needs to have a much lower bandwidth. Hence an outer loop time delay compensation observer is not necessary and good stability of the outer loop is achieved by using a phase lag controller $k(z)$.

In addition to the two feedback loops, a feedforward loop of the utility voltage is also added (with a positive sign) to compensate for the effect of the grid voltage disturbance. The ideal continuous time transfer function of the feedforward loop $F(s)$ can be derived as follows. Assuming for now that the capacitor current observer in Figure 6.2b effectively compensates for sampling time delay in the inner loop, then both the observer block and associated measurements can be removed and the delay blocks can be replaced by unity transfer functions. The PWM block is also assumed to be a unity transfer function. In the continuous time domain, the closed loop transfer function can be shown to be given by [14]:

$$I_2 = \frac{k(s)G_1(s)}{1+k(s)G_1(s)}I_2^* - \frac{G_1(s)}{1+k(s)G_1(s)}(L_1Cs^2 + k_cCs + 1)V_u(s) + \frac{G_1(s)}{1+k(s)G_1(s)}F(s)V_u(s)$$

(6.1)

where $G_1(s)$ is the transfer function of the controlled plant that relates the output current to the inverter voltage, which is given by

$$G_1(s) = \frac{1}{(L_1L_2C)s^3 + (k_cL_2C)s^2 + (L_1 + L_2)s}$$

(6.2)

According to Equation 6.1, to enable the cancellation of the effect of the disturbance V_u, the feedforward loop transfer function should be

$$F(s) = (L_1Cs^2 + k_cCs + 1)$$

(6.3)

The ideal feedforward transfer function involves differentiation of the utility voltage signal, which is undesirable in practice. This was overcome by implementing a slow feedforward loop that measures the steady state harmonics of the utility voltage and calculates the derivative by simply shifting the signal by 90° and multiplying it by $2\pi f$ where f is the harmonic frequency. In fact it was found in practice that compensation for only the fundamental 50 Hz component provides satisfactory results for typical utility THD values [14].

6.3 Digital Sampling Strategy

A DSP was used to implement the controller. In the DSP, a periodic up/down counter is used to represent the symmetrical PWM carrier and a compare unit is used to detect matching between the carrier counter and the modulating signal register. The modulation signal register needs therefore to be updated with the correct value before

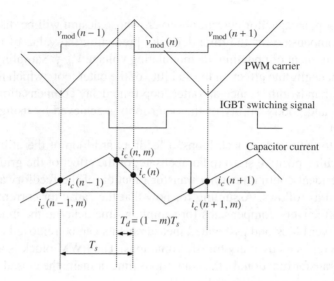

Figure 6.3   PWM carrier and capacitor current

the counter starts counting up or down, which means in effect that the modulating signal register needs to be updated at the peaks and troughs of the PWM carrier signal. However, since a finite time is needed to sample the signal by the DSP and to perform the digital controller calculations, the measured signal needs to be sampled in advance of the update instant, as illustrated in Figure 6.3, that is, there is a time lag between the measurement and control action. Such a time lag can lead to system instability and hence it is essential to take this time lag into account when designing the system, and the inner loop in particular.

In Figure 6.3, the present moment of updating the modulating signal V_{mod} is (n) and the previous moment of update is $(n-1)$. The capacitor current is measured at (n,m), which precedes the update moment (n) by $T_d = (1-m)T_s$ where m is the time lag coefficient and T_s is the sampling period which equals half the period of the PWM carrier, that is, the sampling frequency is double the PWM carrier frequency. Ideally, the time delay T_d needs to be zero $(m=1)$, but in practice, for the DSP used, the maximum value of m was 45%. This is nearly in the middle of the rising or falling edge of the PWM carrier, which is close to the peaks or troughs of the capacitor current and close to the instants of transistor switching and associated switching noise.

We are ideally interested in sampling the mains frequency envelope of the capacitor current, rejecting the high frequency ripple component and switching noise. Achieving this would require sampling near the middle of the rising or the falling edge of the current waveform, that is, near the middle of the on and off periods of the transistors away from the switching instants [22]. To accurately sample the envelope would, however, require a sophisticated algorithm that varies the time delay, which would be expensive computationally. Fortunately, a good approximation would be to sample at the peaks and troughs of the PWM carrier near the ideal sampling instances, which in practice was found to give satisfactory results in terms of ripple and switching noise rejection [20]. However, this means that the time delay needs to be extended to be equal to the full sampling period, which aggravates the instability problem. The stability and transient response of the system can be improved if the time delay is compensated for. This can be achieved by estimating, using an observer, the actual value of the capacitor current at the moment of update using the value measured previously, based on a model of the system. A later section (Section 6.5) discusses the design of such an observer.

6.4 Effect of Time Delay on Stability

As discussed earlier, the outer loop has a relatively low bandwidth and the effect of the time delay is not significant. However, the inner capacitor current feedback loop needs to have a high bandwidth to provide effective damping, and hence the effect of time delay is more critical to the stability of this loop.

To illustrate the effect of time delay on system stability, the root locus of the open transfer function $I_2(z)/E(z)$ (see Figure 6.2b) was plotted with the time delay coefficient m as a variable. Assuming the feedforward loop compensates for the utility disturbance, the utility disturbance and the feedforward loop in Figure 6.2b can be removed from the analysis. Hence, without the time delay and zero order hold (ZOH) blocks, the continuous time domain open loop transfer function is given by

$$\frac{I_2(s)}{E(s)} = \frac{k(s)G_2(s)}{1 + G_2(s)H(s)} \tag{6.4}$$

where

$$G_2(s) = \frac{1}{(L_1 L_2 C)s^3 + (L_1 + L_2)s} \tag{6.5}$$

$$H(s) = k_c L_2 C s^2 \tag{6.6}$$

Including the time delay and ZOH and assuming a proportional controller of $k(s)$, Equation 6.4 can be transformed into the z domain as

$$\frac{I_2(z)}{E(z)} = \frac{kG_2(z, m)}{1 + GH(z)} \tag{6.7}$$

where $G_2(z, m)$ is the modified z-transform of $G_2(s)$ and $GH(z)$ is the z-transform of $G(s)H(s)$ [21]. The root locus of the open loop transfer function in Equation 6.7 with time delay coefficient m as a parameter is shown in Figure 6.4. Instability occurs when the closed loop poles P_3 and P_4 go out of the unit circle, which happens when $m = 20\%$. This means that extending the sampling time delay to a full sampling period will result in an unstable system, unless the effect of time delay is compensated for.

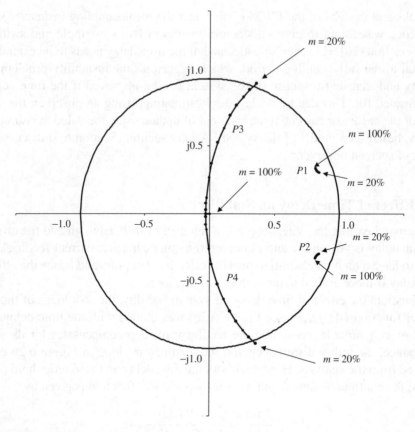

Figure 6.4   Root locus of the system with the variation of the time delay coefficient m

6.5   Capacitor Current Observer

To improve the stability margins, an observer of the inner capacitor current loop can
be utilized to compensate for the time delay by estimating the undelayed value of the
capacitor current, based on a model of the system and the delayed measured value.
A capacitor current observer similar to that proposed by Ito and Kawauchi [23] was
therefore designed as follows. The inner capacitor current loop with the time delay can
be modeled in the z-domain as shown in Figure 6.5. The capacitor current is given by,

$$i_c = i_1 - i_2 \tag{6.8}$$

The transfer function of $\frac{I_1(s)}{V_{L1}(s)}$ including the ZOH but not the time delay is given by:

$$\frac{I_1(s)}{V_{L1}(s)} = g(s) = \frac{1 - e^{-T_s s}}{s} \frac{1}{L_1 s} \tag{6.9}$$

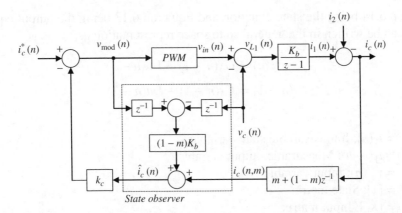

Figure 6.5 Capacitor current loop modified z-transformed with time delay compensation state observer

Time delay less than a full sampling period can be modeled in the z-domain using the modified z-transform [21]. The modified z-transform of Equation 6.9, which is $\frac{I_c(z,m)}{V_{L1}(z,m)}$ can be shown to be given by

$$g(z, m) = \frac{K_b}{z - 1}(m + (1 - m)z^{-1}) \qquad (6.10)$$

where $K_b = T_s/L_1$.

Keeping in mind that the z-transform of $g(s)$ is

$$\frac{I_1(z)}{V_{L1}(z)} = g(z) = \frac{K_b}{z - 1} \qquad (6.11)$$

Then according to Equation 6.10, the delayed current sample is related to the unde-layed current sample via:

$$I_1(z, m) = (m + (1 - m)z^{-1})I_1(z) \qquad (6.12)$$

Taking the inverse z-transform of Equation 6.12 we get

$$i_1(n, m) = i_1(n - 1) + m(i_1(n) - i_1(n - 1)) \qquad (6.13)$$

Taking the inverse z-transform of Equation 6.11 we get

$$i_1(n) = i_1(n - 1) + K_b v_{L_1}(n - 1) \qquad (6.14)$$

Equations 6.13 and 6.14 can be thought of as a state space model relating the delayed measured sample of the current to the synchronously sampled current, with

Equation 6.14 being the state equation and Equation 6.13 being the output equation, which can be written in the general state space representation as:

$$x(n) = Gx(n-1) + Hu(n-1) \tag{6.15}$$

$$y(n-1) = Cx(n-1) + Du(n-1) \tag{6.16}$$

where
$x(n)$ $= i_1(n)$: Immeasurable state variable
$y(n-1) = i_1(n, m)$: Measurable output variable
$u(n)$ $= v_{L_1}(n)$: Input variable
G $= [1]$: State matrix
H $= [K_b]$: Input matrix
C $= [1]$: Output matrix
D $= [mK_b]$: Direct transmission matrix.

A state observer can then be designed to estimate the state based on the measured output such that

$$\hat{x}(n) = G\hat{x}(n-1) + Hu(n-1) + K_e[y(n-1) - \hat{y}(n-1)] \tag{6.17}$$

where K_e is the observer state matrix, $\hat{x}, \hat{y}$ are estimated values of x and y.
From Equation 6.17, the observed current is then given by

$$\hat{i}_1(n) = \hat{i}_1(n-1) + K_b v_{L_1}(n-1) + K_e(i_1(n,m) - \hat{i}_1(n,m)) \tag{6.18}$$

Substituting for $\hat{i}_1(n, m)$ from Equation 6.13 in Equation 6.18 and re-arranging yields

$$\hat{i}_1(n) = \hat{i}_1(n-1) + K_b v_{L_1}(n-1) + K_e(i_1(n,m) - \hat{i}_1(n-1) - mK_b v_{L_1}(n-1)) \tag{6.19}$$

The characteristic equation of the above state observer is given by

$$z - 1 + K_e = 0 \tag{6.20}$$

If the observer pole is located at the origin ($z = 0$), a quick deadbeat response is achieved, when $K_e = 1$, and the observer equation reduces to,

$$\hat{i}_1(n) = i_1(n, m) + (1 - m)K_b v_{L_1}(n-1) \tag{6.21}$$

The above observer simply calculates the value of the current at the moment of update n by linearly extrapolating the measured value of the current, using the derivative obtained from voltage measurement.
Substituting for i_1 from Equation 6.8 into Equation 6.21 and re-arranging yields:

$$\hat{i}_c(n) = i_c(n, m) + (1 - m)K_b v_{L_1}(n-1) + (i_2(n,m) - \hat{i}_2(n)) \tag{6.22}$$

Given the low bandwidth of the i_2, 50 Hz outer loop, the difference between the delayed measured value of i_2 and its value at the moment of update is small. The last term $i_2(n, m) - \hat{i}_2(n)$ in the above equation can therefore be neglected. This was validated by detailed simulation of the system. The capacitor current observer therefore reduces to

$$\hat{i}_c(n) = i_c(n, m) + (1 - m)K_b v_{L_1}(n - 1) \tag{6.23}$$

Neglecting the small voltage drop across the L_2 inductor, a switching ripple-free signal of v_{L_1} can be obtained by subtracting the measured utility voltage from the modulating signal,

$$v_{L_1}(n - 1) = v_{mod}(n - 1) - v_u(n - 1) \tag{6.24}$$

This saves the cost of direct measurement of v_{L_1} and virtually eliminates the PWM ripple component that exists in the actual signal, which was found using detailed simulation to result in the deterioration of the quality of the output current. This observer is illustrated in the block diagram of Figure 6.5.

6.6   Design of Feedback Controllers

The inner loop controller k_c and the outer loop controller $k(z)$ have to be designed to give satisfactory performance in terms of transient response, steady-state response, and disturbance rejection. The first option is to choose proportional controllers. To determine the effect of the loop gains on the transient response of the system, the root locus has been plotted for different values of the inner loop gain k_c as shown in Figure 6.6. As is well known, increasing the output loop gain k will decrease the relative stability by pushing the closed loop poles to the right. However, increasing k_c will improve the transient response by pushing the whole root locus to the left. Hence increasing k_c and decreasing k will improve the transient response by increasing the stability margins and reducing the overshoot and settling time. The loop gains have also to be chosen to increase the system ability to reject the utility harmonic disturbance. As stated earlier, the feedforward loop is implemented by calculating the derivative of the fundamental component of the utility voltage and adding this to the modulating voltage but this does not compensate for the utility harmonic components. Therefore, it is essential to optimize the controller so it can attenuate these harmonics. The disturbance-output transfer function can be obtained by setting $I_2^* = 0$ and $F(s) = 0$ in Equation 6.1, which gives

$$N(s) = \frac{I_2(s)}{V_u(s)} = -\frac{1 + k_c(s)Cs + L_1Cs^2}{1 + k(s)G_1(s)}G_1(s) \tag{6.25}$$

Figure 6.7 shows $N(s)$ for different values of k_c and k. F_o is the frequency at which $\frac{I_2(s)}{V_u(s)} = 0$. NR is the maximum attenuation for frequency below F_o. As k_c increases, F_o decreases, which means decreasing the range of attenuated harmonic frequencies. As k increases, both F_o and NR increase, which means increasing the range of attenuated

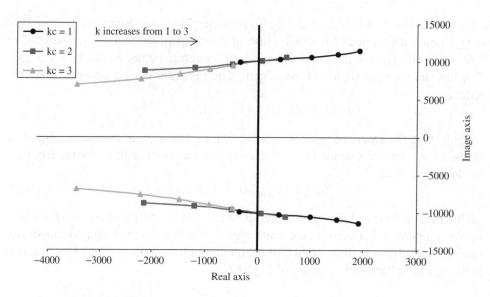

Figure 6.6   Root locus of $G_2(s)$ for different values of inner loop gain k_c

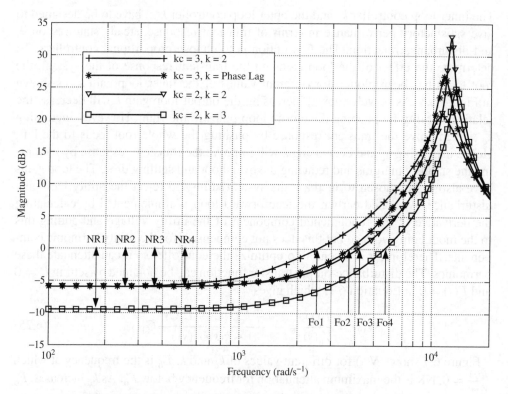

Figure 6.7   Frequency plot of $N(s)$ for different loop controllers

harmonic frequencies as well as increasing the attenuation of these harmonics. As can be seen, decreasing k_c and increasing k improves disturbance rejection, which contradicts the condition of improving stability which is achieved by increasing k_c and decreasing k. Therefore, a compromise has to be made. If both gains are chosen to be 2, the disturbance rejection ratio NR is 5.1 dB and F_o is 560 Hz but the phase margin is only 11°.

It is possible to improve the phase margin of the system without affecting the disturbance rejection performance. This can be achieved by changing $k(z)$ to a phase lag compensator. The design can be carried out in the continuous domain as the outer loop response is slow compared to the sampling rate. The Tustin bilinear transformation is subsequently used to derive a discrete transfer function. The lag compensator is designed to shift the gain crossover frequency slightly to the left and hence increase the phase margin, as shown in Figure 6.8. The pole and zero of the compensator are put as close as possible to the gain crossover frequency so the attenuation caused by the lag compensator does not affect the low frequency range. This is essential to keep the open loop gain high at low frequencies to get maximum utility harmonics rejection. As seen from Figure 6.8 the phase margin has increased from 11° to 31°. The designed phase lag compensator is given by:

$$k(z) = 2\frac{0.7664z - 0.4015}{z - 0.635} \tag{6.26}$$

6.7 Simulation Results

A detailed MATLAB®/Simulink model was used to aid the design and predict the performance of the system under different conditions. This also helped to give further insight into the effect of time delay and controller parameters on stability and output current quality, and to assess the design of the capacitor current observer and validate the assumptions made.

Figure 6.9 shows the sampled capacitor current and the output current when sampling was delayed from the peaks and troughs of the PWM carrier by 70% of T_{sw} ($m = 0.3$), with a utility voltage THD of 0%. The simulation started with the capacitor current observer disabled. At simulation time $t = 0.02$ s, the time delay observer was enabled. The time delay caused poor stability, which can be clearly noticed from the oscillatory response before the observer was enabled. Although, the activation of the time delay observer improved stability and suppressed resonance, distortion can still be observed on the output current due to the sampling of the high switching ripple component.

Figure 6.10 shows the capacitor current and the output current when the time delay was extended to be equal to a full sampling period ($m = 0$) such that sampling

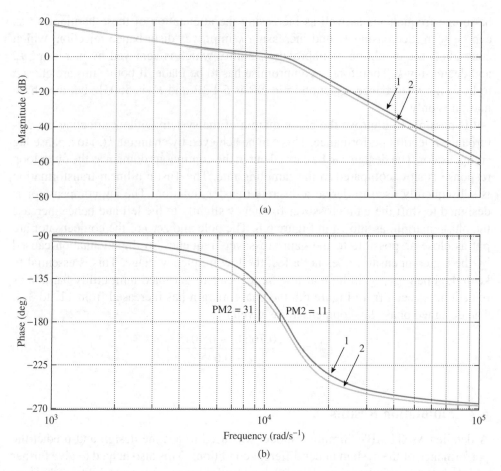

Figure 6.8 Bode diagram without phase lag compensator (a) and with phase lag compensator (b)

took place at the peaks and troughs of the PWM carrier, that is, approximately near the 50 Hz envelope value of the capacitor current, which is also approximately half way between switching instants, as was seen in Figure 6.3. The capacitor current observer was again activated at simulation time $t = 0.02$ s. The extension of the time delay aggravated the instability problem, as can be clearly seen from the strong oscillation appearing on current signals during the first fundamental period. However, when the time delay observer was activated, the system became stable and the capacitor current was virtually ripple free. In addition, the observer did not cause noticeable distortion to the output signal thanks to rejection of the switching ripple. Figure 6.11 shows the demanded and actual grid current for a typical grid voltage THD of 2.0%. The results show that the current THD is 2.6%.

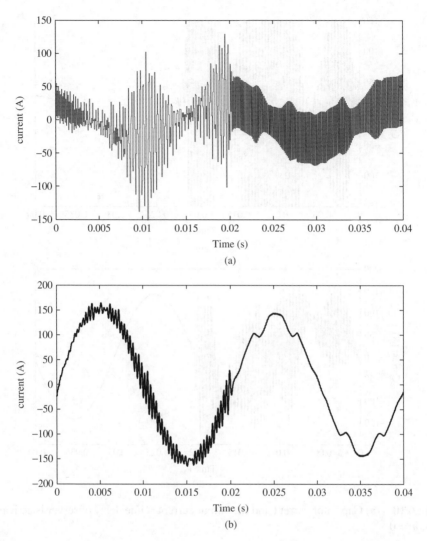

Figure 6.9 (a) Capacitor current and (b) output current. Time delay observer is activated at 0.02 s, $m = 0.3$

The distortion that appears in the output current is due to the presence of utility voltage harmonics.

6.8   Experimental Results

The controller and capacitor current observer described in this chapter were implemented experimentally using TMS320FL2407 DSP. The grid voltage was measured

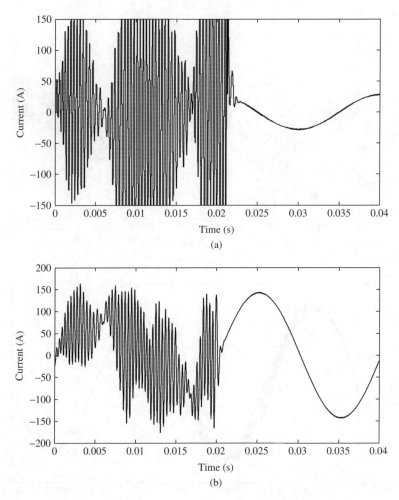

Figure 6.10 (a) Capacitor current and (b) output current. Time delay observer is activated at 0.02 s, $m = 0$

experimentally and its main harmonic components are shown in Figure 6.12, it contains a typical grid voltage harmonics, with a THD of 2.0%. Figure 6.13a shows the sampled capacitor current when the computational time delay coefficient m is set to 0.3. It is clear that this time delay causes the switching ripple component to be sampled. In Figure 6.13b, the sampling time delay is set to T_{sw} and it is evident that this results in the rejection of switching ripple. The distortion that appears on the signals is due to utility voltage harmonics. The spectrum of the capacitor current is presented in Figure 6.14. The experimental output current of 140 A rms is shown in Figure 6.15. The spectrum of the output current is shown in Figure 6.16 and has a THD of 2.8%.

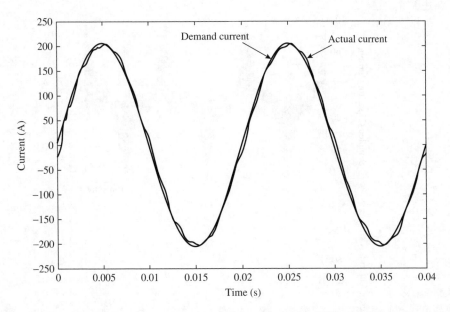

Figure 6.11   Simulated output current and demand signal, THD = 2.6%, utility voltage THD = 2.0%

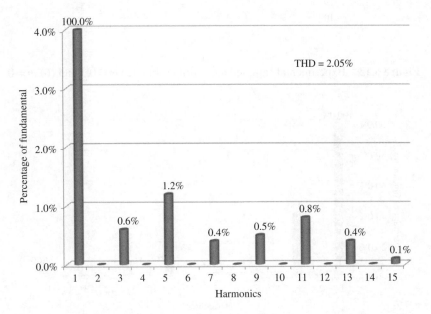

Figure 6.12   Experimental spectrum of grid voltage

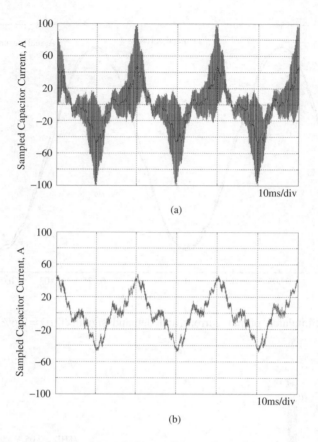

Figure 6.13   Experimental sampled capacitor current (a) $m = 0.3$ and (b) $m = 0$

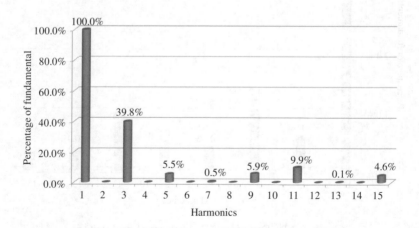

Figure 6.14   Experimental spectrum of the capacitor current

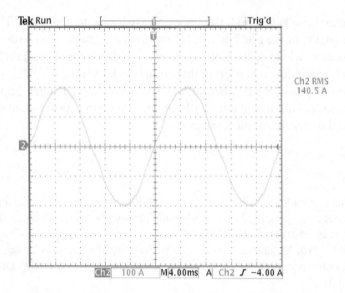

Figure 6.15   Experimental output current, THD = 2.8%, utility voltage THD = 2.0%

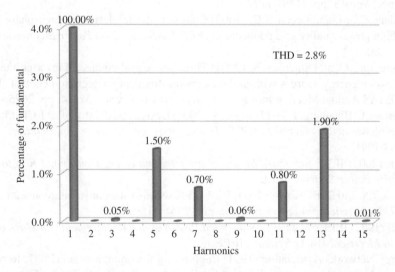

Figure 6.16   Experimental spectrum of output current in Figure 6.15

6.9   Conclusions

The design and practical implementation of a digital current controller for a two-level three-phase PWM voltage source inverter have been presented. Careful selection of the instances of sampling of the capacitor current to occur at the peaks of the PWM away from the switching instances of the transistors and close to the middle of the rising or falling edge of the capacitor current improves the rejection of capacitor current switching ripple and high frequency transistor switching noise. This results in an improved quality of the grid current, but requires the sampling to be delayed by one sampling instance, which reduces the system's stability margin. System stability can be improved by including an observer that compensates for the time delay in the inner capacitor loop. A phase lag compensator in the outer loop has been implemented to further increase the stability margin without compromising steady-state performance and, in particular, grid harmonics rejection.

References

1. Sharkh, S.M., Arnold, R.J., Kohler, J. *et al.* (2006) Can microgrids make a major contribution to UK energy supply? *Renewable and Sustainable Energy Reviews*, **10**, 78–127.
2. Lessetter, R.H. Microgrids. IEEE PES Winter Meeting, Vol. 1, pp. 305–308, 2002.
3. Jaywarna, N., Wu, X., Zhang, Y. *et al.* (2006) Stability of a microgrid. The 3rd IET International Conference on Power Electronics Machines and Drives (PEMD), Dublin, Ireland, pp. 316–320.
4. Twining, E. and Holmes, D.G. (2002) Grid current regulation of a three-phase voltage source inverter with an LCL filter. IEEE 33rd Annual Power Electronics Specialists Conference, Vol. 13, pp. 1189–1194.
5. Prodanovic, M. and Green, T.C. (2003) Control and filter design of three-phase inverters for high power quality grid connection. *IEEE Transactions on Power Electronics*, **18** (1), 373–380.
6. Chongming, Q. and Smedley, K. (2001) Three-phase grid-connected inverter interface for alternative energy sources with unified constant-frequency integration control. The 36th IEEE IAS Annual Meeting Industry Applications Conference, Vol. 4, pp. 2675–2682.
7. Liserre, M., Blaabjerg, F. and Hansen, S. (2005) Design and control of an LCL-filter-based three-phase active rectifier. *IEEE Transactions on Industry Applications*, **41** (5), 1281–1291.
8. IEEE (2003) IEEE Std 1574. *Standard for Interconnecting Distributed Resources with Electric Power Systems*, IEEE.
9. Basso, T.S. and DeBlasio, R. (2004) IEEE 1547 series of standards: interconnection issues. *IEEE Transactions on Power Electronics*, **19** (5), 1159–1162.
10. IEEE (1992) IEEE Std 519*Recommended Practices and Requirements for Harmonic control in Electrical Power Systems*, IEEE.
11. Energy Networks Association (1991) Engineering Recommendation G59/1: Recommendations for the Connection of Embedded Generating Plant to the Public Electricity Suppliers' Distribution Systems.
12. Energy Networks Association (1995) Engineering Technical Report No. 113 Revision 1: Notes of Guidance for the Protection of Embedded Generating Plant Up to 5 MW for Operation in Parallel with Public Electricity Suppliers' Distribution Systems.
13. Lindgrin, M. and Svenson, J. (1998) Control of a voltage-source converter connected to the grid through an LCL-filter – application to active filtering. IEEE 29th Annual Power Electronics Specialist Conference, Vol. 1, pp. 229–235.
14. Sharkh, S.M. and Abu-Sara, M.A. (2004) Current control of utility-connected two-level and three-level PWM inverters. *EPE Journal*, **14** (4), 13–18.

15. Sharkh, S.M. and Abu-Sara, M.A. (2001) Current control of utility-connected DC-AC three-phase voltage source inverters using repetitive feedback. 9th European Power Electronics Conference, Graz, Austria.
16. Sharkh, S.M., Hussien, Z.F., and Sykulski, J.K. (2000) Current control of three-phase PWM inverters for embedded generators. IEE 8th International Conference on Power Electronics and Variable Speed Drives, London, UK, pp. 524–529.
17. Abdul-Rahim, N.M. and Quaicoe, J.E. (1994) Analysis and design of a feedback loop control strategy for three-phase voltage-source utility interface. Conference records of the 1994 IEEE Industry Applications Society Annual Meeting, Vol. 2, pp. 895–902.
18. Blasko, V. and Kaura, V. (1997) A novel control to actively damp resonance in input LC filter of a three-phase voltage source converter. *IEEE Transactions on Industry Applications*, **33** (2), 542–550.
19. Hussien, Z.F. (2000) Control of three-phase PWM inverter for flywheel energy storage systems. PhD Thesis. University of Southampton.
20. Abu-Sara, M.A. (2004) Digital control of utility and parallel connected three-phase PWM inverters. PhD Thesis. University of Southampton.
21. Ogata, K. (1995) *Discrete-time Control Systems*, Prentice-Hall.
22. Van de Sype, D.M., Gusseme, K.D., Van de Bossche, A.P. and Melkebeek, J.A.A. (2004) A sampling algorithm for digitally controlled boost PFC converter. *IEEE Transactions on Power Electronics*, **19** (3), 649–657.
23. Ito, Y. and Kawauchi, S. (1995) Microprocessor-based robust digital control for UPS with three-phase PWM inverter. *IEEE Transactions on Power Electronics*, **10** (2), 196–203.

7

Design and Control of a Grid-Connected Interleaved Inverter

7.1 Introduction

Most commercially available three-phase grid-connected inverters are based on the two-level voltage source PWM (pulse width modulation) bridge inverter topology with an LCL output filter [1–5], similar to that described in the previous chapter. The switching frequency of these inverters tends to decrease as the power level increases, limited mainly by switching losses. As a result, high power inverters tend to have disproportionately large filter components. Besides the obvious disadvantage of large size and high cost, large filter components have other drawbacks on the system. First, using large inductors slows down the dynamic response of the system, which can be critical in cases of fault ride-through during grid disturbance, such as voltage sag and swell. Secondly, large capacitors draw high currents, which in effect reduce the system's output power factor unless control action is taken to compensate for this, for example, by using a feedforward loop, as explained in [6] and Chapter 6. Thirdly, large capacitors provide an easy path for harmonic currents caused by grid voltage harmonics which increase the output current total harmonic distortion (THD). Fourthly, a grid-side inductor is sometimes necessary in a classical two-level bridge inverter to block the high frequency ripple current from being injected into the grid, but that adds further to size and cost. Furthermore, if the system is to be used in a microgrid, where it is required to operate in grid-connected mode as well as in a standalone mode, this additional inductor would be undesirable as it increases the output voltage THD significantly when supplying a nonlinear load.

The need to improve the efficiency and reduce the size and cost of both the inverter and the output filter encouraged more research into using different inverter topologies. A three-level inverter topology has been shown to halve the output inductor

Power Electronic Converters for Microgrids, First Edition. Suleiman M. Sharkh, Mohammad A. Abusara, Georgios I. Orfanoudakis and Babar Hussain.
© 2014 John Wiley & Sons, Ltd. Published 2014 by John Wiley & Sons, Ltd.
Companion Website: www.wiley.com/go/sharkh

ripple current for a given switching frequency, thus reducing the size of the inductor. Additionally, the power switches in a three-level inverter have half the voltage rating of power switches in a two-level inverter, and hence can have faster switching frequencies, thus enabling further reduction in filter size [6–8].

In this chapter we discuss an alternative topology for grid-connected inverters, namely an interleaved topology, as shown in Figure 7.1. The multiphase interleaved topology has been recently gaining popularity, especially in DC/DC converters [9–19] for different power applications, such as electric and hybrid electric vehicles [9–11], communication power supplies [12], power factor corrections for small handheld tools [13, 14], and power boost circuits for PV (photovoltaic) systems [16, 18]. The main motives for using this topology in the aforementioned applications are to increase power density, enhance dynamic performance, and increase efficiency. In addition, the interleaved topology has been used in DC/AC [20, 21] and AC/DC [22–24] converter applications. Interleaving is a form of paralleling technique where a single converter channel, for example, half a bridge with an output inductor, is replaced by N smaller channels connected in parallel, with their switching instants phase shifted equally over a switching period. There are many advantages for the interleaved topology over the conventional topology when used in grid-connected inverters. By introducing a phase shift between the switching instants of the parallel channels of each phase, the amplitude of the total ripple current is N times less and its frequency is N times greater than that of a conventional inverter [22]. The reduction in total current ripple amplitude and the increase in its frequency reduce the size of the required filter capacitance considerably. Furthermore, sharing the current among a number of channels enables the use of smaller, lower current power switches which can switch at high frequency, thus allowing a reduction in inductor size. The net result is that the size of the filter and the overall size of the system are smaller than an equivalent classical two-level bridge inverter with LCL output filter. The reduction in total current ripple and the high frequency also eliminate the need for a second output filter inductor (i.e., using an LC filter instead of an LCL filter) that is used in two-level and multi-level grid-connected inverters to block the switching ripple in cases where the grid impedance is too low. Additionally, the resonance frequency of the interleaved system is high due to the much smaller output filter capacitor and equivalent inductor. This improves the dynamic response of the system and its ability to ride through grid disturbances, such as voltage sags and swells. It also gives more headroom for increasing the controller gain at lower harmonic frequencies, using phase lag compensation, for example, in order to suppress the low harmonic currents caused by grid voltage harmonic distortion. Finally, the switching losses are spread over several components, with greater overall surface area, which improves cooling considerably. Although replacing one channel by several requires more gate drives and more current sensors, the improvements in terms of size reduction, speed of response, and improved output current THD were found to justify the extra complexity, and accordingly the system described in this chapter is now produced commercially at a competitive price.

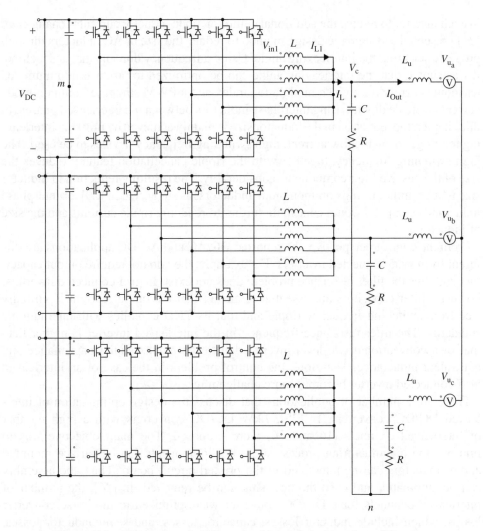

Figure 7.1   Three-phase interleaved grid-connected interleaved inverter

Although the use of the interleaved topology has been investigated for different DC/AC and AC/DC applications, it has not been explored in depth in the context of grid-connected inverters. Asiminoaei *et al.* [20] proposed an interleaved topology to parallel multi-active power filters on the AC bus to increase power capability and to reduce AC line inductors. It was shown that although interleaving reduces the resultant switching harmonic current due to ripple cancellation, it introduces additional common mode currents in the three phases (compared to a single inverter) due to current flowing from one interleaved inverter to another when the inverters' space vector voltages are simultaneously zero. In order to keep the benefit of the interleaving topology in terms of reducing inductance value, common mode inductors are used in each inverter to reduce the additional common mode inductor currents. Zhang *et al.* [21] presented a systematic design method to reduce the size of AC inductors through properly selecting the interleaving angle. Using a frequency domain analysis method, it was shown that the interleaving angle can be optimized to produce minimum AC harmonic current for a given modulation index and a PWM strategy. Common mode currents were dealt with using interphase inductors between the interleaved phases. In this chapter, the reduction in the inductor size is mainly achieved by utilizing interleaving topology to enable low current, high switching frequency devices to be used. The high switching frequency, together with the ripple cancellation feature, reduces the size of the passive filter components considerably. The resulting reduction in inductor size is substantial. Using common mode inductors, as suggested in [20], or interphase inductors, as in [21], would contribute further to reducing ripple currents and the size of inductors.

From the control perspective, the reported DC/AC and AC/DC applications are different from grid-connected inverters. For example, they do not require output capacitors and thus the filter resonance problem does not exist. In grid-connected inverters, however, filter capacitors are essential to ensure that the output current is virtually free from switching frequency ripple and that its THD complies with international standards. The filter resonance frequency in the interleaved inverter is higher than that of a conventional two-level inverter due to the much smaller capacitance and equivalent inductance. Therefore, the control problem in the case of an interleaved grid-connected inverter becomes more challenging.

There are a number of publications that discuss the design optimization of interleaved DC/DC converters. In [25], 1 kW DC/DC converters with a high number of interleaved channels (16 and 36) were proposed. The main objective was to improve the manufacturing process so power components can be surface mounted and/or inductors can be integrated in the printed circuit board. Thus, the assembly can be automatic, and even the heat sink can be removed. In [26], the number of interleaved channels for a DC/DC converter was optimized to minimize converter losses, which include inductor losses, capacitor losses, and semiconductor losses. Oliver *et al.* [27] discussed the effect of the number of interleaved channels on the size of the filter inductors and capacitors. However, in all the above publications the switching frequency was fixed in the design procedure and hence it was not used as an optimization parameter. Nussbaumer *et al.* [14] proposed design guidelines of a boost DC/DC converter that take into account the number of channels, the switching frequency, and the converter operating mode to achieve minimal size.

This chapter describes a novel three-phase interleaved grid connected inverter with LC filter. The interleaved topology enables low current high frequency devices to be used which, together with the ripple cancellation feature, reduces the overall size of the inverter and filter by about 50% compared to an equivalent classical two-level inverter. A systematic and detailed procedure to select the number of channels, the switching frequency, and the filter components to achieve minimum inductor size and satisfactory harmonics rejection is proposed. The design and practical implementation of the control system are also discussed in detail. Section 7.2 is devoted to ripple cancellation in an interleaved inverter. In Section 7.3 the hardware design is discussed. The controller structure is described in Section 7.4. In Section 7.5 a model of the system is derived. In Section 7.6 the design of the control system is discussed, and in Section 7.7 simulation and practical results are presented.

7.2 Ripple Cancellation

The output of each interleaved channel (Figure 7.1) is connected to a point of common coupling (PCC) to the grid through an inductor L, carrying a share I_{Lx} of the total current I_L such that:

$$I_L = \sum_{x=1}^{N} I_{Lx} \tag{7.1}$$

The common points of the inductors in each phase are connected to star-connected capacitors, C, in series with resistors, R. The voltage sources V_u and the inductances L_u (with subscripts a, b, and c for the three phases) represent the grid equivalent circuit, and I_{Out} is the grid current.

To understand the ripple cancellation mechanism, consider an interleaved converter with $N = 3$. The three switching signals S_1, S_2, and S_3 are phase shifted by $T_{sw}/3$, as shown in Figure 7.2. The individual inductor currents are also shown. When the three currents are added together at the point where the inductors are connected to each other, they form the total inductor current I_L. If the "on" state is defined as the state when the upper switch is on, the peak-to-peak ripple in the total inductor current I_L

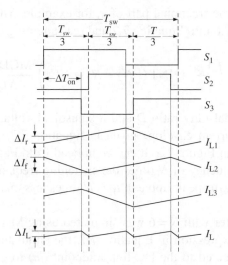

Figure 7.2 Switching signals and inductor ripple currents in an interleaved inverter with three channels

can be expressed, based on Figure 7.2, as

$$\Delta I_L = K_{on}\Delta I_r - (N - K_{on})\Delta I_f \tag{7.2}$$

where K_{on} is the number of "on" interleaved channels and is given by:

$$K_{on} = INT(DN) \tag{7.3}$$

where D is the duty ratio, $INT(x)$ gives the smallest following integer of x, for example, $INT(2.1) = 3$. ΔI_r is the channel peak-to-peak rising current during ΔT_{on} and is given by:

$$\Delta I_r = \frac{\left(\frac{V_{DC}}{2} - V_c\right)\Delta T_{on}}{L} \tag{7.4}$$

ΔI_f is the channel peak-to-peak falling current during ΔT_{on} and is given by:

$$\Delta I_f = \frac{\left(\frac{V_{DC}}{2} + V_c\right)\Delta T_{on}}{L} \tag{7.5}$$

The time ΔT_{on} is given by:

$$\Delta T_{on} = \frac{MOD(DN)}{Nf_{sw}} \tag{7.6}$$

where $MOD(x)$ gives the fractional part of x, for example, $MOD(2.1) = 0.1$. Substituting Equations 7.3–7.6 into Equation 7.2 and rearranging gives

$$\Delta I_L = \left(\frac{V_{DC}}{2}(2INT(DN) - N) - NV_c\right)\frac{MOD(DN)}{Nf_{sw}L} \tag{7.7}$$

Assuming a sinusoidal duty ratio D and a sinusoidal voltage V_c at the PCC, the peak-to-peak ripple current envelope given in Equation 7.7 is plotted for different values of N as shown in Figure 7.3. It can be noticed from Figure 7.2 that the ripple current in I_L has a frequency of N times the switching frequency. Figure 7.3 shows that the maximum peak-to-peak ripple in I_L is N times less than that of the individual channels.

An interleaved inverter with $N = 6$ was simulated using MATLAB®/Simulink. To verify the analytical expression in Equation 7.7, the simulated inverter also has its neutral point (n) connected to the DC-link midpoint (m) to give three independent phases so the inverter output voltage is $\pm V_{DC}/2$. Figure 7.4 shows the capacitor current which validates the results given in Equation 7.7 and Figure 7.3.

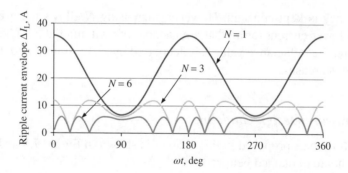

Figure 7.3   Peak-to-peak ripple current envelope

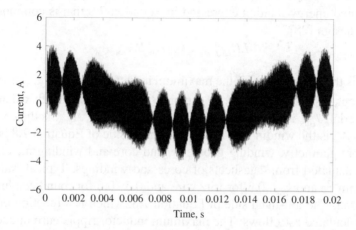

Figure 7.4   Simulated capacitor current with $N=6$, the neutral point (n) is assumed to be connected to the DC-link midpoint (m)

7.3   Hardware Design

The hardware design of a grid connected interleaved inverter involves the selection of the switching frequency f_{sw}, the number of interleaved channels N, the channel inductance L, inductor core and coil, and filter capacitance C. The main optimization criterion used in this chapter aims to achieve the smallest inductor size as it significantly contributes to the inverter's size and cost. The main design constraints taken into account are: first, the switching frequency is limited by the maximum allowed losses in the semiconductors. Secondly, the number of interleaved channels is limited by the maximum number of PWM outputs of the DSP (digital signal processor).

As discussed in the previous section, the individual inductor current is inversely proportional to N. To keep the ratio of the switching ripple to the fundamental component constant, the inductance has to increase linearly with N. Knowing that the inductor size is proportional to LI^2 [14, 28], the total inductor size remains the same as N increases, if the frequency is kept constant. However, increasing N allows the use of smaller IGBTs with lower current rating that are capable of switching at a higher frequency and, therefore, the inductance value L is decreased by increasing N, and thus the total inductor size decreases.

7.3.1 Hardware Design Guidelines

The design flowchart proposed in this chapter is shown in Figure 7.5 and consists of four steps that are explained below:

Step 1 For the whole range of switching frequencies and number of channels considered in the design, select a suitable inductor core (from a core database) according to the stored energy relation expressed in Equation 7.8 that is commonly used in inductor design [28],

$$L\widehat{I}I_{Lx,rms} < k_{Cu}J_{rms}\widehat{B}A_{core}A_w \tag{7.8}$$

where L is the inductance, $\widehat{I}$ is the maximum inductor current, $I_{Lx,rms}$ is the inductor rms current, k_{Cu} is the copper fill factor, $\widehat{B}$ is the maximum flux density of the core material, J_{rms} is the current density in the winding, A_{core} is the core sectional area, and A_w is the winding area. The right-hand side of Equation 7.8 is a function of the core geometry, winding geometry, and core and winding materials, which can be calculated from datasheets of cores and windings. Typical values for J_{rms} ($2-4$ A mm^{-2}) and k_{Cu} (0.3 for Litz wire and 0.5–0.6 for round conductors) [28], can be used. The left-hand side of Equation 7.8, which is a function of N and f_{sw}, can be calculated as follows: The maximum inductor ripple current occurs during the zero crossing of the grid voltage (during which the duty ratio is 50%), thus the inductance required to limit the ripple to the maximum value of ΔI_{Lx} is given by:

$$L = \frac{V_{DC}}{4f_{sw}\Delta I_{Lx}} \tag{7.9}$$

Equation 7.9 ignores the phase interaction between the three phases and assumes the inverter voltage to be $\pm V_{DC}/2$. This is an acceptable assumption when $N > 1$ during the zero crossing of the grid voltage. To explain this, consider a two-channel interleaved inverter, as depicted in Figure 7.6. It consists of two conventional two-level half bridge inverters with the filter inductors parallel to each other. The grid impedance is ignored and hence the grid voltage is connected to the paralleling points of the inverters. The three phase modulating signals D_a, D_b, and D_c and the PWM signals during the grid zero crossing of phase (a) are shown in Figure 7.7. The switching interval T_{sw} is divided into six periods T_1 to T_6. During each period, different inverters' switching states are applied. Using the principle of superposition, the utility voltage can be set to zero and the voltage across the inductor due to the inverter voltage can be calculated.

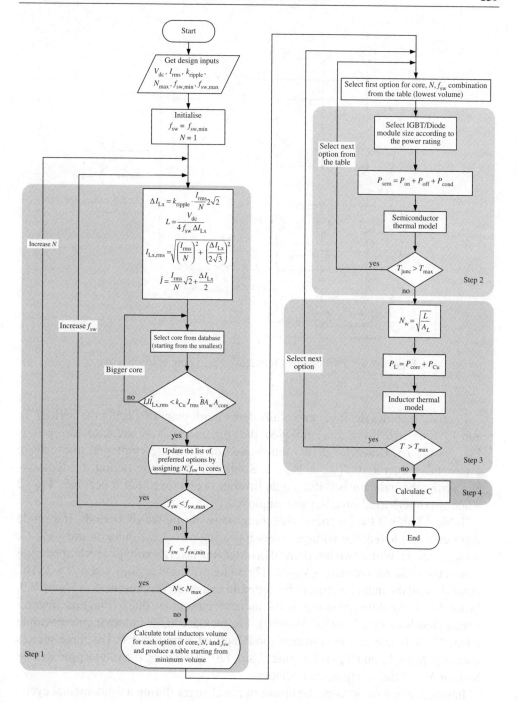

Figure 7.5 Hardware design flowchart

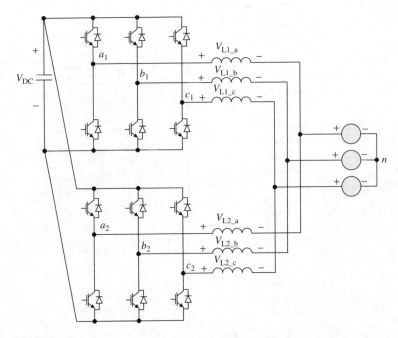

Figure 7.6   Two-channel interleaved grid-connected inverter

Figure 7.8 shows the inverter's equivalent circuit diagram during the periods T_3 and T_4. If a single inverter is employed, the voltage across the inductor during T_3 is $V_{L1\_a} = V_{DC}/3$. However, if two interleaved inverters are employed, the voltage across the inductor is $V_{L1\_a} = V_{DC}/2$. Similarly, during the period T_4, the voltage across the inductor is 0 if a single inverter is employed but it becomes $V_{DC}/2$ when two interleaved inverters are employed.

Table 7.1 shows the inverters' switching states during the six periods. If a single inverter is employed, the voltage levels appearing across the inductor are $+V_{DC}/3$, $-V_{DC}/3$, and 0. With a two interleaved inverters the inverter voltage levels appearing across the inductor are only $\pm V_{DC}/2$. The same principle applies when $N > 2$. The simulation of the inductor current for different values of N is presented in Figure 7.9. When $N = 1$, the different slopes in the inductor current are due to the three inverter voltage levels ($+V_{DC}/3$, $-V_{DC}/3$, and 0). The increase in the inductor current ripple when $N > 1$ is due to the common mode current introduced in the three phases during periods T_1 and T_4 (see Figure 7.8b), [20]. However, the total ripple will be less for $N > 1$ due to ripple cancellation.

In practice, the peak-to-peak current ripple changes during a fundamental cycle, but for the purpose of estimation of the inductor size the ripple may be assumed to be constant. Assuming a constant ratio k_{ripple} of the inductor peak-to-peak ripple

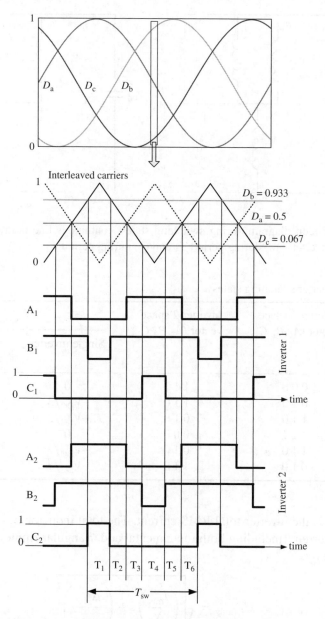

Figure 7.7 Switching signals during zero crossing of grid voltage

current ΔI_{Lx} to the peak-to-peak fundamental current, ΔI_{Lx} is given by:

$$\Delta I_{Lx} = k_{\text{ripple}} \cdot \frac{I_{\text{rms}}}{N} 2\sqrt{2} \qquad (7.10)$$

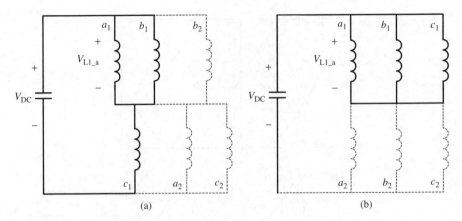

Figure 7.8 Circuit diagram during switching states, continuous line is inverter 1 and the dotted line is inverter 2 (a) during period T_3 and (b) during period T_4

Table 7.1 Inverter switching states

Period	Inverter 1 space vector $(A_1\ B_1\ C_1)$	Inverter 2 space vector $(A_2\ B_2\ C_2)$	V_{L1_a}	
			Single inverter	Two interleaved inverters
T_1	0 0 0	1 1 1	0	$-V_{DC}/2$
T_2	0 1 0	1 1 0	$-V_{DC}/3$	$-V_{DC}/2$
T_3	1 1 0	0 1 0	$+V_{DC}/3$	$+V_{DC}/2$
T_4	1 1 1	0 0 0	0	$+V_{DC}/2$
T_5	1 1 0	0 1 0	$+V_{DC}/3$	$+V_{DC}/2$
T_6	0 1 0	1 1 0	$-V_{DC}/3$	$-V_{DC}/2$

where I_{rms} is the inverter total RMS current. The total inductor rms current $I_{Lx,rms}$ and peak current (including both fundamental and triangular ripple) can be shown to be given by:

$$I_{Lx,rms} = \sqrt{\left(\frac{I_{rms}}{N}\right)^2 + \left(\frac{\Delta I_{Lx}}{2\sqrt{3}}\right)^2} \qquad (7.11)$$

$$\hat{I} = \frac{I_{rms}}{N}\sqrt{2} + \frac{\Delta I_{Lx}}{2} \qquad (7.12)$$

If the inequality of Equation 7.8 does not hold then a bigger size inductor core is required. Using datasheets of inductor cores, calculate the total volume of inductors per combination option of N and f_{sw}. Create a list of preferred options starting from the one with minimum inductor volume.

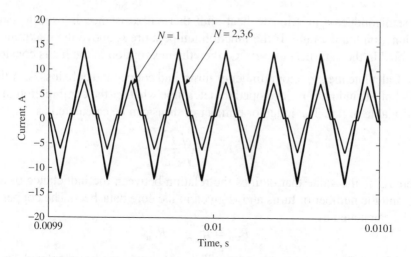

Figure 7.9 Simulation of the inductor current during the zero crossing of the grid voltage, $V_{DC} = 700V, f_{sw} = 35\text{kHz}$, and $L = 190\mu H$

Step 2 Select suitable IGBT/diode modules and calculate the semiconductor losses for each option of the list created in step 1. A thermal model is used to determine whether this option is viable or not. The semiconductor power losses per IGBT/diode module are given by:

$$P_{sem} = P_{on} + P_{off} + P_{cond} \tag{7.13}$$

where P_{on} is the turn-on losses, P_{off} is the turn-off losses, and P_{con} is the conduction losses. Because silicon carbide (SiC) diodes are used in this design, reverse recovery diode losses are neglected. The losses in Equation 7.13 can be calculated according to the semiconductor module datasheet such as

$$P_{on} = \frac{f_{sw}}{T} \int_0^T \frac{V_{DC}}{V_{base}} E_{on}(I_{Lx\_on}(\omega t))dt \tag{7.14}$$

$$P_{off} = \frac{f_{sw}}{T} \int_0^T \frac{V_{DC}}{V_{base}} E_{off}(I_{Lx\_off}(\omega t))dt \tag{7.15}$$

$$P_{cond} = I_{T,\text{rms}}^2 \cdot R_{CE} + I_{T,\text{rms}} \cdot V_{CE} + I_{D,\text{rms}}^2 \cdot R_{BE} \tag{7.16}$$

where T is the fundamental cycle period, V_{base} is the base voltage given in the datasheet, and E_{on}, E_{off} are the turn-on and turn-off energy losses per switch. I_{Lx_on}, I_{Lx_off} are the turn-on and turn-off inductor currents, respectively. $I_{T,\text{rms}}$ and $I_{D,\text{rms}}$ are the IGBT and diode RMS currents which can be derived using circuit simulation or calculation [13]. R_{CE} and R_{BE} are the IGBT and diode on resistances and V_{CE} is the IGBT forward voltage drop. After calculating the semiconductor power losses using Equation 7.13, a thermal model that calculates the semiconductor junction to heat sink thermal resistance is used to calculate the junction temperature. If the junction temperature is above the maximum limit specified in the datasheet ($\sim 120\,°C$) then the next option in the list is considered.

Step 3 Calculate the inductor number of turns, and copper and core losses. A thermal model of the inductor is developed to determine whether the combination of N and f_{sw} is viable or not. The inductor number of turns can be calculated as

$$N_w = \sqrt{\frac{L}{A_L}} \tag{7.17}$$

where A_L is the value that defines the relation between the inductance of a given core and the number of turns and is given in the core datasheet. The copper losses can be calculated as

$$P_{Cu} = I_{Lx,rms}^2 \cdot R_{Cu} \tag{7.18}$$

where R_{Cu} is the windings resistance. The core losses can be calculated according to the Steinmetz equation [29] as

$$P_{core} = k \cdot f_{sw}^a \cdot \widehat{B}^b \tag{7.19}$$

where the parameters k, a, and b can be obtained from the core datasheets. The next step is to develop a thermal model of the inductor to check whether or not the internal temperature exceeds the maximum allowed limit defined in the datasheets of the winding and core. The thermal model of the inductor will not be discussed in this chapter. In [14], detailed guidelines for developing such a model are provided. The development of the thermal model depends on the inductor cooling method. If the thermal model shows that that inductor hotspot temperature is higher than the maximum limit, then the next option in the list is considered.

Step 4 Calculate the required filter capacitance. The value of the filter capacitance C is determined so that the switching frequency component in the output current I_{Out} is attenuated by the required percentage r. The value of the resistor R is selected to provide the required system damping and will be discussed in Section 7.5. However, this resistor will affect the attenuation of the filter so the effect of R needs to be taken into account. Due to the interleaved topology, the high frequency current in the output current I_{Out} has a frequency of Nf_{sw}. Therefore, the required attenuation ratio of the high frequency component with respect to the fundamental is given by:

$$r = \frac{I_{Out,Nf_{sw}}}{I_{Out,f_o}} \tag{7.20}$$

Using superposition, the response for each frequency component can be considered separately, and the grid can be assumed to be a short circuit when we consider the effect of the high switching frequency component only [30], therefore by analyzing the resulting equivalent circuit, Equation 7.20 can be written as

$$r = r_1 \cdot r_2 \tag{7.21}$$

Table 7.2   Epcos cores data

Core	A_{core} (mm^2)	A_w (mm^2)	V_T (mm^3)
PM 62/49	270	570	62 000
PM 74/59	442	790	101 000
PM 87/70	657	910	133 000
PM 114/93	1 070	1 720	344 000

where

$$r_1 = \frac{I_{L,Nf_{sw}}}{I_{Out,f_o}} = \frac{V_{L,f_{sw}}}{N\omega_{sw}LI_{Out,f_o}} \tag{7.22}$$

$$r_2 = \frac{I_{Out,Nf_{sw}}}{I_{L,Nf_{sw}}} = \sqrt{\frac{1 + N^2\omega_{sw}^2R^2C^2}{(1 - N^2\omega_{sw}^2CL_u)^2 + N^2\omega_{sw}^2R^2C^2}} \tag{7.23}$$

where $\omega_{sw} = 2\pi f_{sw}$ and $V_{L,f_{sw}}$ is the RMS inductor voltage component at the switching frequency. Given that N, f_{sw}, and L have all been determined in step1, C is selected to give the required current attenuation r.

7.3.2   Application of the Design Guidelines

In this section, the design guidelines developed above will be applied to design a grid-connected interleaved inverter. The inputs for the design are $V_{DC} = 700$ V, $I_{rms} = I_{Out,f_o} = 85$ A, $f_{sw,min} = 10$ kHz, $f_{sw,max} = 50$ kHz, $k_{ripple} = 0.7$, and $N_{max} = 6$. The maximum number of channels is restricted by the maximum number of PWM outputs of the DSP that will be used in the experimental implementation, which is TMS320F2808. Four cores made of the ferrite N27 material produced by Epcos [31] have been considered. These cores can be easily fixed on a heat sink for better cooling. Inductors can be encased in the heat sink as was proposed in [9]. This approach, however, increases losses due to eddy current generated in the heat sink by fringing flux and thus it was not considered. The core and winding areas and the total volume of these cores are summarized in Table 7.2.

Step 1 For the whole range of f_{sw} and N, the inductor stored energy $L\hat{I}I_{Lx,rms}$ has been calculated and compared with the core size factor ($k_{Cu}J_{rms}\hat{B}A_{core}A_w$) to select the proper core according to Equation 7.8. The parameters used in calculating the cores size factor are: $k_{Cu} = 0.6$, $J_{rms} = 3.5$ Amm^{-2}, and $\hat{B} = 290$ mT which is 65% of the saturated flux density of the ferrite material N27. Figure 7.10 shows how the total inductor volume changes versus switching frequency. The discontinuities in the curves happen when the next size core is selected. Table 7.3 summarizes the best 10 options of Figure 7.10. The comments in Table 7.3 are based on the results of Steps 2 and 3 below.

Step 2 It is quite obvious that the first two options in Table 7.3 require relatively high switching frequencies if we take into account the amount of current per channel. Therefore, unless SiC switches are used, conventional power switches are not capable of operating at this combination of current and switching frequency. In addition, Option a (one channel) and option b (two interleaved channels) will not give the required performance in terms of ripple cancellation and hence filter capacitance reduction. For the rest of the options in Table 7.3, it is possible to find semiconductor components that operate at the required current rating and switching frequency and without the junction temperature exceeding the maximum limit.

Step 3 Options c, d, and e did not pass this step as the thermal models showed that the inductor winding hotspot temperature can exceed the maximum temperature of 140 °C. Option f consists of only one channel and the interleaving benefits are lost. Option g, which consists of six interleaved channels with a minimum switching frequency of 30 kHz, satisfies the inductor thermal requirements and hence it has been chosen. The switching frequency was increased to 33 kHz so it is slightly higher than the minimum limit of 30 kHz. The inductance value for this design according to Equation 7.9 is $L = 190 \,\mu H$. Figure 7.11 shows the results of the inductor power losses for the PM62/49 core for different design points. The solid line represents the maximum allowed power dissipation so the hotspot winding temperature does not exceed the maximum temperature of 140 °C. Increasing the switching frequency decreases the ripple current and thus the inductor copper losses but it also increases the core losses. However, the decrease in copper losses is much more than the increase in core losses due to the use of ferrite cores.

Step 4 Given that $L = 190 \,\mu H$, an attenuation ratio of $r_1 = 0.024$ is obtained according to Equation 7.22. By choosing $C = 15 \,\mu F$, and in the presence of a small grid impedance of only 3 μH, an attenuation ratio of $r_2 = 0.135$ is obtained ($R = 0.5 \,\Omega$). Therefore the total attenuation ratio will be $r = 0.3\%$, which provides the required attenuation stated in [32]. This requires the presence of a grid-side impedance of at least 3 μH to provide the required attenuation which can be guaranteed by the EMI filter used at the output of the inverter.

7.4 Controller Structure

In the practical implementation of this system, one DSP is used per phase. To avoid any unnecessary communication between the controllers, the three phases of the inverters are controlled independently. Therefore the dq approach [3] is not appropriate. Figure 7.12 shows the block diagram of the PWM inverter, LC filter, damping resistor R, and the control system of one of the phases. The model assumes balanced

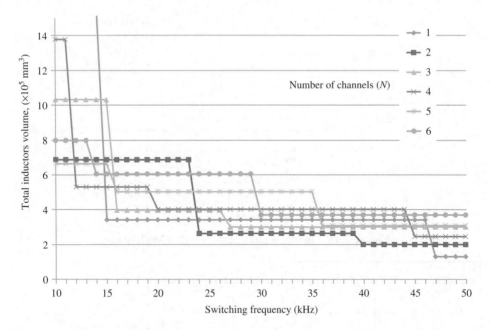

Figure 7.10   Total volume of inductors versus switching frequency

Table 7.3   Options for minimum inductor volume

Option	$\frac{I_{rms}}{N}$ (A)	Minimum f_{sw} (kHz)	Core	Total inductor volume (mm³)	Comments
a	1	85.0	47	PM 87/70	133 000	High switching losses and no interleaving benefit
b	2	42.5	40	PM 74/59	202 000	High switching losses
c	4	21.3	45	PM 62/49	248 000	High inductor losses
d	3	28.3	27	PM 74/59	303 000	High inductor losses
e	5	17.0	36	PM 62/49	310 000	High inductor losses
f	1	85.0	15	PM 114/93	344 000	No interleaving benefit
g	6	14.2	30	PM 62/49	372 000	Best choice
h	4	21.3	20	PM 74/59	404 000	Large inductor size
i	5	17.0	16	PM 74/59	505 000	Large inductor size
j	4	21.3	12	PM 87/70	532 000	Large inductor size

three-phase currents and hence the voltage of the start point of the filter capacitors is approximately at the same potential as the grid neutral point. The current in each inductor is controlled using a single feedback loop with a digital controller $K(z)$. Each interleaved channel is equipped with a Hall-effect sensor measuring the induc-

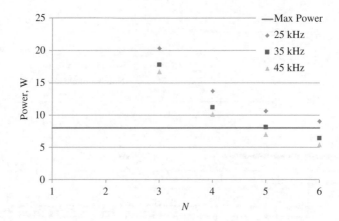

Figure 7.11   Inductor power losses with PM62/49 core

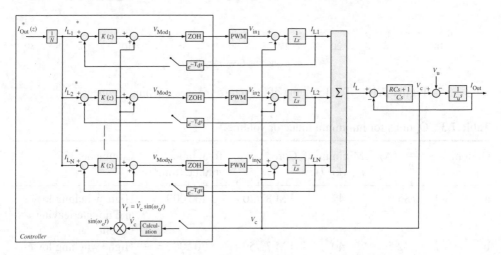

Figure 7.12   Block diagram of one phase and its controller

tor current. There are two important reasons for using six sensors to measure the individual currents instead of one sensor measuring the total inductor current as was reported in [19]. First, current sharing between channels can be monitored and an alarm signal can be issued if one channel is not operating properly. Secondly, and more importantly, individual sensors can ensure proper protection of the switching devices against short circuit current faults. If one sensor is used to measure the total current, an individual current can reach the maximum limit before the total current reaches the total maximum limit. Also, the speed of detection of high current is very critical in this application because, in the interleaved topology, the rate of change of the current (di/dt) can be much higher than that in an equivalent conventional low switching frequency topology. This makes protection against short circuit current

Table 7.4 Natural damped frequency versus L_u

L_u (µH)	Natural resonance frequency f_n (kHz)
5	19.6
20	11.4
100	7.8

more challenging. The problem is further aggravated when inductors with ferrite cores are used, as they tend to saturate quickly, which leads to a rapid reduction of their inductance and a further increase in the di/dt.

A feedforward loop of the grid voltage at the PCC V_c is also included to cancel the grid voltage disturbance at the fundamental frequency. A second feedback loop of the capacitor current to provide active damping, as discussed in [5], was not implemented as the sampling frequency f_s (33 kHz) is only 1.7–4.2 times the natural resonance frequency f_n, depending on the value of the grid inductance, as shown in Table 7.4, which is too low to provide effective damping. In order to control the system resonance frequency, the sampling rate needs to be at least 8–10 times faster than the natural resonance frequency [33]. Instead, a resistor in series with the filter capacitor is used to provide passive damping. Fortunately, due to the ripple cancellation feature of the interleaved topology, the capacitor current is quite small and hence the power dissipation in R is also small and the losses are acceptable.

The time delay T_d caused by the controlling processor computational time is modeled as $e^{-T_d s}$. The total inductor current I_L equals the sum of the capacitor current and the output current I_{Out}, and hence the demanded I_L^* should ideally include a correction to allow for the capacitor fundamental frequency 50 Hz current. But since this current is very small in practice, it may be neglected and the demanded I_L^* can be set to the value of the required output current. In this set-up, the input DC link is regulated by an external boost circuit to 700 V DC so there is no need to take into account the DC voltage variation.

7.5 System Analysis

In this section, the effects of the passive damping resistor R, computational time delay T_d, and the variation in grid impedance L_u on system stability are studied. The system's ability to reject grid harmonics is also analyzed. For the purpose of the discussion in this section, which is meant to provide an insight into the problem, the controller $K(z)$ is assumed to be a simple proportional gain with a value of 10. In the next section, the design of a more sophisticated controller $K(z)$ is discussed.

From Equation 7.1 and Figure 7.12, the system can be described by the following three equations:

$$I_{\text{Out}} = \frac{1}{L_u s}(V_c - V_u) \tag{7.24}$$

$$V_c = \frac{RC + 1}{Cs}(I_L - I_{\text{Out}}) \tag{7.25}$$

$$I_L = \frac{N}{Ls}(V_{\text{in}_x} - V_c) \tag{7.26}$$

From Equations 7.24–7.26, the inductor current in any channel I_{Lx} is given by:

$$I_{Lx} = (V_{\text{in}_x} - B(s)V_u)A(s) \tag{7.27}$$

where

$$A(s) = \frac{L_u Cs^2 + RCs + 1}{LL_u Cs^3 + RC(L + NL_u)s^2 + (L + NL_u)s} \tag{7.28}$$

$$B(s) = \frac{RCs + 1}{L_u Cs^2 + RCs + 1} \tag{7.29}$$

$A(s)$ is the open loop transfer function of I_{Lx} to V_{in_x} and $B(s)$ is the grid disturbance transfer function of I_{Lx} to V_u. A simplified block diagram of the system is shown in Figure 7.13. The continuous time domain open loop transfer function including the computational time delay is therefore given by:

$$G(s) = e^{-T_d s}A(s) \tag{7.30}$$

In the discrete time domain, $G(z)$ can be obtained by performing the z-transform of $G(s)$ taking into account the zero order hold effect,

$$G(z) = Z\left[\frac{1 - e^{-T_s s}}{s}e^{-T_d s}A(s)\right] \tag{7.31}$$

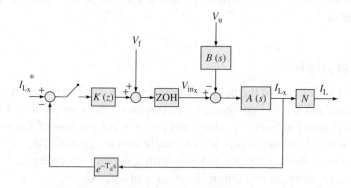

Figure 7.13   Single channel equivalent block diagram

$G(z)$ was computed using MATLAB® with the time delay $e^{-T_d s}$ approximated using the Pade approximation.

7.5.1   Effect of Passive Damping and Grid Impedance

Without computational time delay ($T_d = 0$) and with R set to zero, the root locus of $K(z)G(z)$ is shown in Figure 7.14, with L_u as a parameter varying from 1 to 500 µH. The closed loop poles are located at the border of the unit circle, which means the system will be critically stable. In Figure 7.15, the root locus of $K(z)G(z)$ is plotted when R is set to 0.5 Ω. It is clear that the closed loop poles have been pushed well inside the unit circle, making the system more stable. However, the system seems to suffer from a lack of immunity to grid impedance variations as it becomes unstable when $L_u > 20$ µH.

7.5.2   Effect of Computational Time Delay

The effect of the computational time delay on system stability was discussed in chapter 6 an it has been discussed in the literature by many researchers and linear time delay observers have been proposed [34–36]. In this system, such an observer is not necessary, as will be shown in the following discussion. Figure 7.16 illustrates the sampling strategy for the proposed controller. The processor updates the modulating

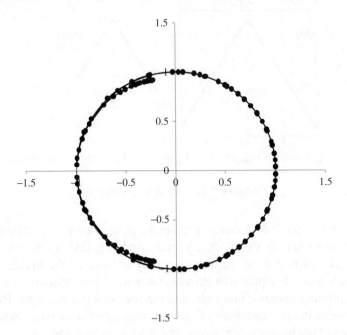

Figure 7.14   Root locus of $K(z)G(z)$ with L_u changing from 1 to 500 µH, $T_d = 0$, $K(z) = 10$, $R = 0$ Ω

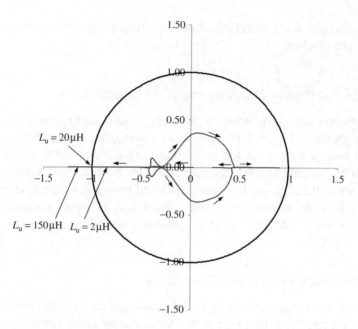

Figure 7.15 Root locus of $K(z)G(z)$ with L_u changing from 1 to 500 μH, $T_d = 0$, $K(z) = 10$, $R = 0.5\,\Omega$

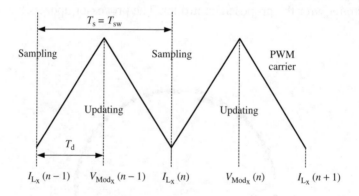

Figure 7.16 Sampling strategy

voltage V_{Mod} when the PWM carrier reaches its peak. Due to the finite time needed by the A/D converter to sample the current and the DSP to perform controller calculations, the current needs to be sampled in advance of the update moment. In order to sample a nearly ripple free current component, it is better to extend this time so that the sampling instance takes place at the previous trough of the PWM carrier, away from the switching instances of the transistors and associated switching noise as was discussed in chapter 6. Therefore, the time delay is set to

$$T_d = 0.5T_s \tag{7.32}$$

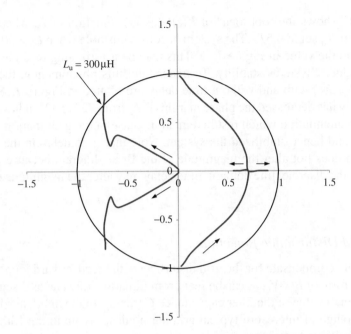

Figure 7.17 Root locus of $K(z)G(z)$ with L_u changing from 1 to 500 µH, $T_d = 0.5T_s$, $K(z) = 10$, $R = 0.5\,\Omega$

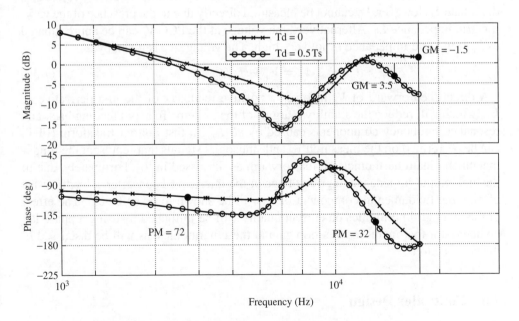

Figure 7.18 Bode diagram of $K(z)G(z)$ with and without time delay, $L_u = 40$ µH

Figure 7.17 shows the root locus of $K(z)G(z)$ as a function of L_u when R is set to 0.5 Ω but with T_d set to $0.5T_s$. The system becomes unstable when $L_u > 300\,\mu H$ which is higher than the value in Figure 7.15. This may be a surprising result as time delay normally reduces systems stability. To understand this phenomenon, the Bode diagrams of $K(z)G(z)$ with and without time delay are plotted in Figure 7.18. Although the time delay has decreased the phase margin (PM) from 72° to 32°, it has also caused a magnitude attenuation which resulted in an increase in the gain margin (GM) from −1.5 to 3.5 and hence stabilized the system. Note that time delay in the continuous time domain does not alter the magnitude of the Bode diagram because $|e^{-T_d s}| = 1$. However, in the discrete time domain, time delay may alter the magnitude of the Bode diagram.

7.5.3 Grid Disturbance Rejection

To completely compensate for the grid disturbance, the feedforward loop should ideally take the form of $B(s)V_u$, as can be seen from Equation 7.27 and as discussed in [6]. Due to the small value of the filter capacitance C (thanks to the interleaved topology), $B(s)$, in the range of interest of typical grid harmonics, say up to the 13th harmonic, can be approximated as

$$B(s) \approx 1 \tag{7.33}$$

Therefore, a direct feedforward of the grid voltage V_u should effectively reject grid disturbance. However, V_u cannot be measured directly due to the presence of the varying grid impedance L_u. Alternatively, the voltage at the PCC V_c can be approximated as V_u if the grid impedance is not very high. The voltage V_c is given by:

$$V_c = V_u + sL_u I_{Out} \tag{7.34}$$

A direct feedforward of V_c involves a positive feedback of the first derivative of I_{Out} which will reduce the stability margins of the system. To avoid feeding back any resonance frequency components caused by $sL_u I_{Out}$, a fast Fourier transform (FFT) can be performed on the measured V_c and sinusoidal signals that represent the fundamental and main harmonics are fed forward, as discussed in [6]. Fortunately, due to the high output impedance to grid voltage of the interleaved converter, a feedforward of only the fundamental component of V_c will be shown to be sufficient. The effect of the main grid harmonics on current quality will be mitigated by using a phase lag compensator that increases the loop gain at these harmonics, as will be discussed in the next section.

7.6 Controller Design

The controller $K(z)$ needs to be designed to fulfill certain tasks. First, it should provide good tracking of the reference signal. Secondly, it should provide good rejection of grid voltage harmonics. Thirdly, the controller stability has to be immune to variations in grid impedance. Good tracking and good grid harmonics rejection require a

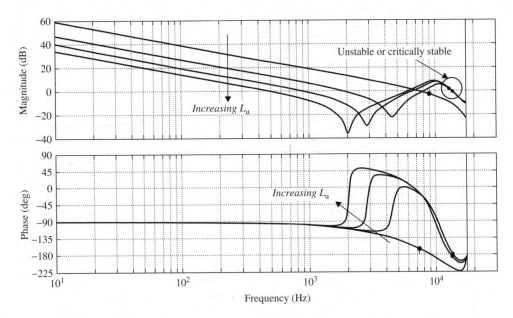

Figure 7.19   Bode diagram of $K(z)G(z)$, $K(z) = 10$, $L_u = 5$, 50, 250, and 500 µH

high gain value of $K(z)$ at the fundamental and the low frequency harmonics. However, a high value of $K(z)$ means less stability. For $K(z) = 10$, the system will become
unstable when $L_u > 300$ µH, as was shown in Figure 7.17. Figure 7.19 shows the Bode
diagram of $K(z)G(z)$ with L_u having the values of 5, 50, 250, and 500 µH. At high values of L_u, the GM becomes negative and hence the system becomes unstable. If $K(z)$
is modified to be a phase lag compensator, as illustrated in Figure 7.20 then it will provide more attenuation at the higher frequencies to improve the GM and at the same
time increase the gain at the lower frequencies to improve reference signal tracking
and grid harmonics rejection. The phase lag compensator has been designed in the
equivalent continuous domain. The real pole and zero of the phase lag are located
at 160 and 1850 Hz, respectively. The gain of the compensator has been determined
to be 4.31, so the overall compensated system has minimum phase and GMs of 28°
and 8, respectively for the whole range of L_u considered. The Bode diagram magnitude plot (see Figure 5.20) crosses the 20 dB line (which is equivalent to $K(z) = 10$) at
around $f = 900$ Hz. This means that the phase lag compensator increases the gain for
$f < 900$ Hz (to improve signal tracking and grid harmonics rejection), and provides
attenuation for $f > 900$ (to improve stability) compared to a proportional controller
with $K(z) = 10$. It is interesting to note that the high phase-crossover frequency of the
uncompensated system (~ 10 kHz depending on L_u) which is a result of the high bandwidth of the interleaved topology (smaller C and equivalent L/N) allows the use of a
simple phase lag compensator as an effective method for improving the gain at the low
frequency harmonics and at the same time improving the system stability by lowering

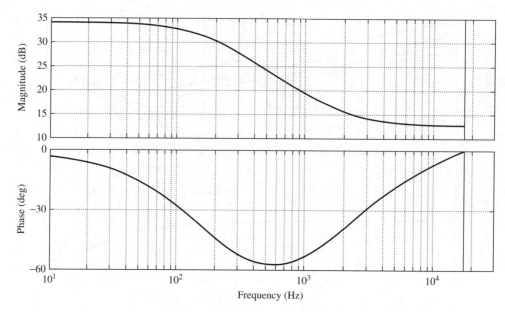

Figure 7.20   Bode diagram of a phase lag $K(z)$, $K(z) = 10\frac{0.5z-0.35}{z-0.97}$

the high frequency gain in order to improve the GM. This, for example, was not possible for the system reported by the authors in [5] as the phase-crossover frequency was quite low ($\sim 2\,\text{kHz}$) (see also Chapter 6).

Having determined the pole, zero and gain of the lag compensator, it can be expressed as

$$K(s) = 4.31\frac{s + 2\pi \times 1850}{s + 2\pi \times 160} \tag{7.35}$$

For the practical implementation, the designed phase lag has been discretized using the bilinear Tustin transformation such as $z = \frac{1+(T_s/2)s}{1-(T_s/2)s}$ and is given by:

$$K(z) = 10\frac{0.5z - 0.35}{z - 0.97} \tag{7.36}$$

Figure 7.21 shows the Bode diagrams of $K(z)G(z)$ with $K(z)$ as given in Equation 7.36 for different values of L_u. It can be noticed that the system always has a positive GM. Also, the low frequency gain is now higher, which will improve reference signal tracking and grid harmonics rejection. The root locus of $K(z)G(z)$ with L_u varying from 0 to 1 mH is shown in Figure 7.22. The system remains always stable and the immunity of the system to grid impedance variation is clear. The effect of $K(z)$ on grid disturbance rejection can be studied by considering the closed loop transfer function of I_L to V_u. By neglecting the feedforward loop, sampling effect,

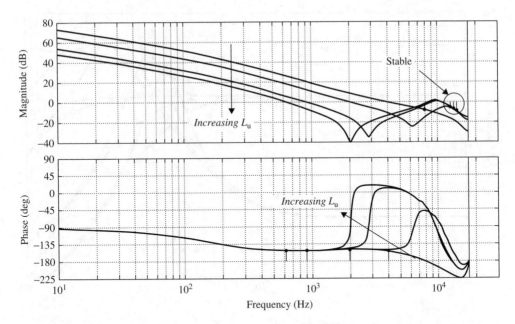

Figure 7.21   Bode diagram of $K(z)G(z)$, $K(z) = 10\frac{0.5z-0.35}{z-0.97}$, $L_u = 5$, 50, 250, and 500 μH

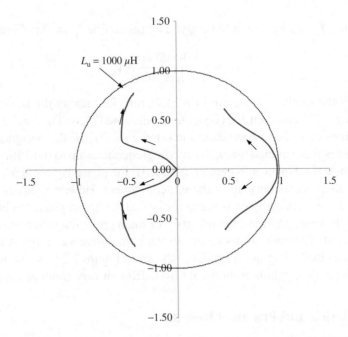

Figure 7.22   Root locus of $K(z)G(z)$ with L_u changing from 0 to 1 mH, $T_d = 0.5T_s$, $R = 0.5\,\Omega$, and $K(z) = 10\frac{0.5z-0.35}{z-0.97}$

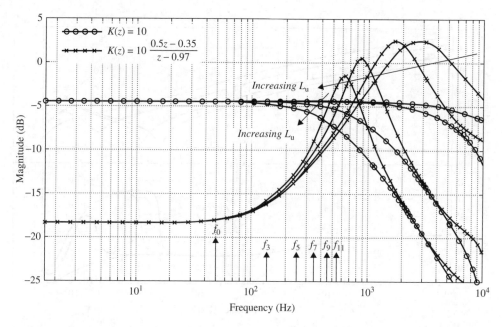

Figure 7.23 Bode diagram of I_L/V_u with $K(z) = 10$, $K(z) = 10\frac{0.5z-0.35}{z-0.97}$ $L_u = 5$, 50, 250, and 500 µH

and time delay, I_L can be shown to be given by (assuming $I_{Lx}^* = 0$ in Figure 7.13)

$$I_L = \frac{-A(s)B(s)}{1 + K(s)A(s)}NV_u \tag{7.37}$$

where $K(s)$ is the Laplace transform of $K(z)$. Figure 7.23 shows the Bode diagram of I_L/V_u with $K(z) = 10$ and with $K(z)$ equal to the designed phase lag. For each case four different values of L_u have been examined. For $L_u < 50\,\mu H$, the designed phase lag always provides more attenuation at harmonic frequencies up to the 15th component. In the extreme case, however, where $L_u > 500\,\mu H$, the phase lag provides attenuation at harmonic frequencies only up to the 9th component. However, this is an extreme case of high L_u and normally L_u is much lower and hence the phase lag compensator will, most of the time, provide attenuation of the main grid voltage harmonics up to the 13th component. The controller's ability to track the reference signal is tested using the closed loop Bode diagram $I_L(z)/I_L^*(z)$ shown in Figure 7.24. At the fundamental frequency, the bode magnitude shows almost 0 dB with very small phase angle shift.

7.7 Simulation and Practical Results

A detailed MATLAB®/Simulink model was used to aid the design and predict the performance of the system. The system parameters used in the simulation are listed in Table 7.5. The simulated grid voltage included low frequency harmonics similar to

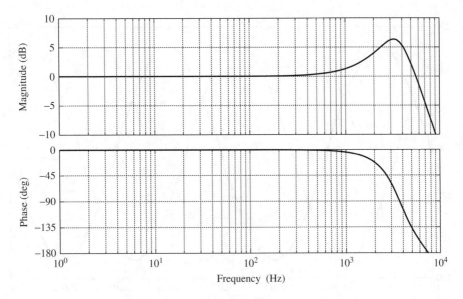

Figure 7.24 Closed loop Bode diagram of $\frac{I_L(z)}{I_L^*(z)}$

Table 7.5 System parameter values

Description	Symbol	Value
Number of interleaved channels	N	6
Passive damping resistor	R	$0.5\,\Omega$
Channel inductor	L	$190\,\mu\text{H}$
Filter capacitor	C	$15\,\mu\text{F}$
Switching frequency	f_{sw}	33 kHz
Sampling frequency	f_s	33 kHz
Time delay	T_d	$15.15\,\mu\text{s}$
Grid voltage	V_u	230 V (rms)
Grid frequency	f_o	50 Hz
Inverter dc voltage	V_{DC}	700 V DC
Inverter nominal power	P_o	60 kW
Inverter nominal output current	$I_{\text{Out\_o}}$	87 A (rms)

those measured at the test site. The total voltage THD was 2.2%. Figure 7.25 shows the three-phase output currents for an 80 A (rms) demand. The current THD was only 2.4%. Figure 7.26 shows the filter capacitor current. It is slightly different from the result presented in Figure 7.4 because of the phase interaction between the three phases due to the fact that the neutral (n) is not connected to the DC-link midpoint (m). From Figure 7.26, the power dissipation in the passive resistor can be calculated to be less than 1.5 W. As the system nominal power is 60 kW, the effect of the passive

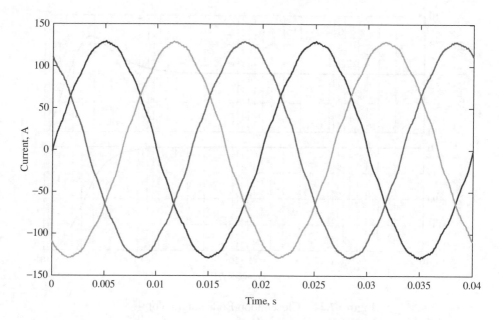

Figure 7.25 Simulation of the output current

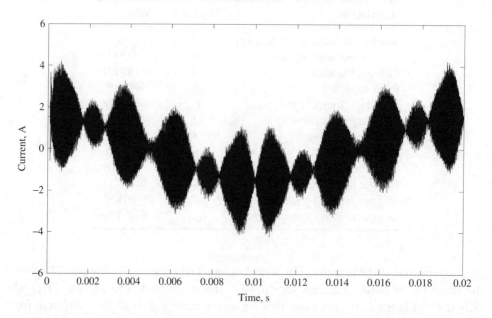

Figure 7.26 Simulated capacitor current of the proposed system (floating DC-link midpoint)

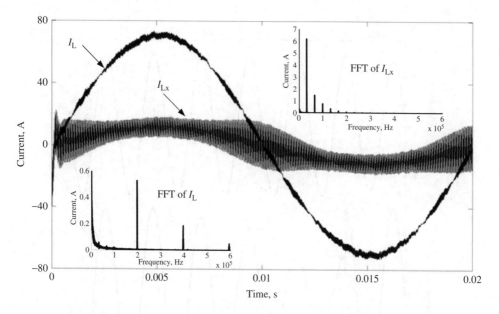

Figure 7.27 Simulated channel inductor current and total inductor current

resistor power dissipation on the system efficiency is basically negligible. Figure 7.27 shows one channel current I_{Lx}, total inductor current I_L, and their harmonic spectra. It is quite clear that the ripple in I_L has been greatly attenuated with its frequency increased by N times the switching frequency. It can be noticed that the I_{Lx} can be negative during the positive cycles and positive during the negative cycles. This is due to the high ratio of the ripple peak-to-peak value to the fundamental peak-to-peak value (k_{ripple}), which is quite acceptable in the interleaved topology as most of the ripple will be canceled out. The dynamic system response to grid swell is shown in Figure 7.28. The grid voltage has a sudden swell of 10% at 0.03 s. The system shows good dynamic response to grid disturbances. The feedforward loop is gradually adjusted according to the new measurement of grid voltage and consequently the current follows its reference more closely within a few cycles.

A three-phase grid-connected interleaved inverter with LC filter was designed and built using the same parameters listed in Table 7.5. To give an insight into the size and cost benefits gained from using the interleaved topology, a size comparison between the proposed inverter and the two-level inverter reported by the authors in [6] is illustrated in Figure 7.29, which shows a single phase of the proposed interleaved inverter and the filter inductor in the inverter reported in [6]. Overall, the interleaved inverter is approximately half the size of the inverter in [6].

The proposed controller was implemented using the Texas Instrument TMS320F 2808 32-bit DSP. This processor has the capability of generating six interleaved PWM outputs. The internal counter of the first PWM carrier is set to give the required switching frequency. The second PWM counter is synchronized with the first counter and delayed by $T_s/6$ (T_s is the sampling period). The third PWM counter is synchronized

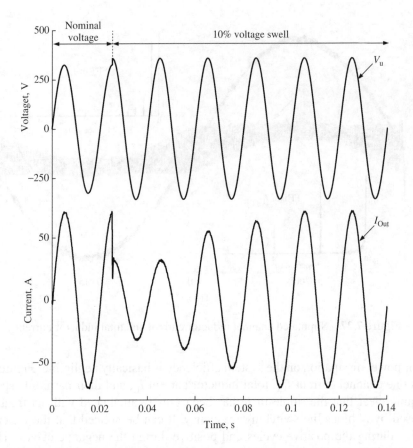

Figure 7.28   Controller response to grid voltage swell ($L_u = 50\,\mu\text{H}$), (simulation results)

with the second counter and delayed by $T_s/6$, and so on. This produces six interleaved triangular carriers to generate carrier-based PWM outputs. One DSP per phase was used and the low speed communications between the controllers, such as start/stop and total current commands, were implemented using the controller area network (CAN) protocol. Synchronization with the grid was implemented by having each phase controller measuring its corresponding grid phase voltage to detect the zero crossing. The reference sine waves were generated internally by the individual controllers using look-up tables of 660 samples (33 kHz/50 Hz). The sine waves amplitude is set externally (by a setting in the user interface) and sent via CAN-Bus to the three phase controllers. The input DC is regulated by an external boost circuit to 700 V DC, as mentioned earlier.

Figure 7.30 shows the three phase output currents. Three inductor currents from one of the phases are shown in Figures 7.31 and 7.32 for two different time scales. The balancing between the channels is confirmed. Figure 7.33 shows the grid voltage and its spectrum as measured at the PCC; the THD is 2.3%. Figure 7.34 shows the output current and its spectrum. The current THD was measured to be 2.8%. Figure 7.35

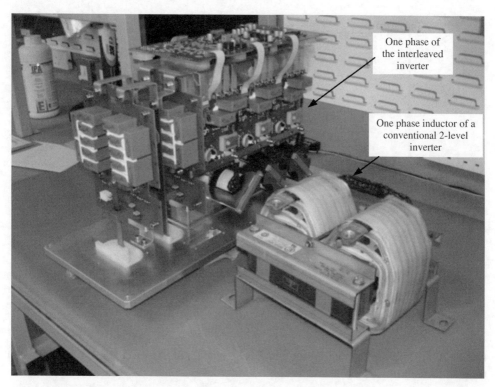

Figure 7.29 Size comparison between one of the proposed inverter phases and the inductor used in the inverter reported in [6]

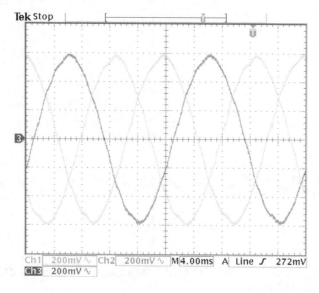

Figure 7.30 Three phase output currents (1 A/5 mV, experimental results)

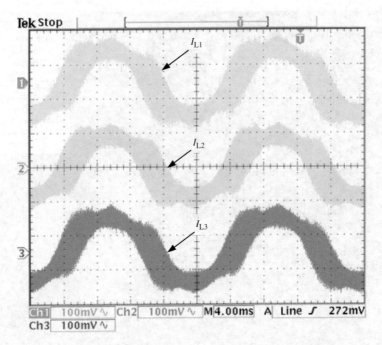

Figure 7.31 Inductor interleaved currents I_{L1}, I_{L2}, and I_{L3} of phase a (1 A/5 mV, experimental results)

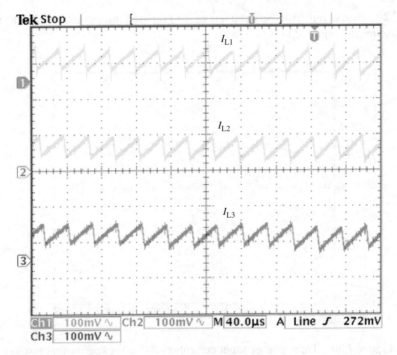

Figure 7.32 Inductor interleaved currents (1 A/5 mV, experimental results)

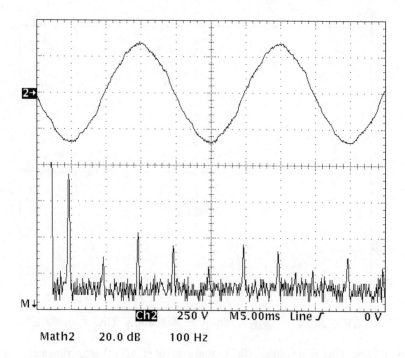

Math2   20.0 dB   100 Hz

Figure 7.33   Grid voltage and its spectrum (experimental results)

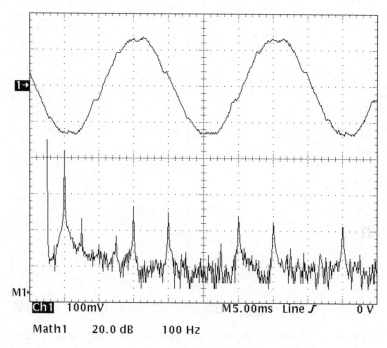

Math1   20.0 dB   100 Hz

Figure 7.34   Output current and its spectrum (1 A/2 mV, experimental results)

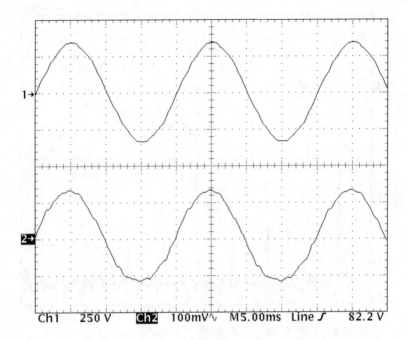

Figure 7.35   Ch1 grid voltage, Ch2 output current (1 A/2 mV, experimental results)

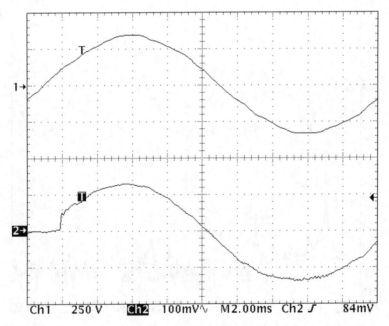

Figure 7.36   Output current step response. Ch1 grid voltage, Ch2 output current (1 A/2 mV, experimental results)

shows the grid voltage and the inverter output current when the grid voltage THD was measured to be only 1.8%. The output current THD was measured to be 2.3%. Figure 7.36 shows the step response of the output current when the demand changes from zero to full value. Good transient response is observed. The inverter has also been tested with various output inductors up to 1 mH, to represent the grid impedance, and robust stability has been confirmed.

7.8 Conclusions

The design and control of a three-phase voltage source grid-connected interleaved inverter with LC filter has been presented. The interleaved topology offers the advantages of reduced filter size, and high grid disturbance rejection compared to an equivalent conventional two-level voltage source inverter with an LCL output filter. The number of channels and the switching frequency have been optimized to provide minimum inductor size, within constraints of thermal limits of the inductor and semiconductor devices, and the maximum frequency possible. The high resonance frequency of the filter requires passive damping resistors in series with the filter capacitors, which was found to be acceptable due to the low capacitor currents. The grid impedance variations were found to reduce system stability. However, using a phase lag compensator incorporated in the inductor current loop was found to be sufficient to increase the system's immunity to grid impedance variations. The design has been validated by simulation and practical results.

References

1. Sharkh, S.M., Abu-Sara, M., and Hussien, Z.F. (2001) Current control of utility-connected DC-AC three-phase voltage-source inverters using repetitive feedback. Proceedings of the 9th EPE, August 27–29, p. 6.
2. Twining, E. and Holmes, D.G. (2002) Grid current regulation of a three-phase voltage source inverter with an LCL filter. Proceedings of the 20th Annual IEEE PESC, Jun. 23–27, Vol. 13, pp. 1189–1194.
3. Prodanovic, M. and Green, T.C. (2003) Control and filter design of three-phase inverters for high power quality grid connection. *IEEE Transactions on Power Electronics*, **18** (1), 373–380.
4. Shen, G., Xu, D., Cao, L. and Zhu, X. (2008) An improved control Strategy for grid-connected voltage source inverters with an LCL filter. *IEEE Transactions on Power Electronics*, **23** (4), 1899–1906.
5. Abusara, M.A. and Sharkh, S.M. (2011) Digital control of a three-phase grid connected inverter. *International Journal of Power Electronics*, **3** (3), 299–319.
6. Sharkh, S.M. and Abu-Sara, M. (2004) Digital current control of utility connected two-level and three-level PWM voltage source inverters. *European Power Electronic Journal*, **14** (4), 13–18.
7. Bouhali, O., B. Francois, C. Saudemont, and E.M. Berkouk (2006) Practical power control design of a NPC multilevel inverter for grid connection of a renewable energy plant

based on a FESS and a Wind generator. Proceedings of the 32nd Annual IEEE IECON, November 7–10, pp. 4291–4296.

8. Araujo, S.V., Engler, A., Sahan, B., and Antunes, F. (2007) LCL filter design for grid-connected NPC inverters in offshore wind turbines. Proceedings of the 7th IEEE ICPE, October 22–26, pp. 1133–1138.

9. Gerber, M., Ferreira, J.A., Hofsajer, I.W. and Seliger, N. (2005) High density packaging of the passive components in an automotive DC/DC converter. *IEEE Transactions on Power Electronics*, **20** (2), 268–274.

10. Destraz, B., Louvrier, Y., and Rufer, A. (2006) High efficient interleaved multi-channel DC/DC converter dedicated to mobile applications. IEEE Industry Applications Conference, pp. 2518–2523.

11. Pavlovsky, M., Tsuruta, Y., and Kawamura, A. (2008) Pursuing high power-density and high efficiency in DC-DC converters for automotive application. Proceedings of the IEEE PESC, June 15–19, pp. 4142–4148.

12. Li, W. and He, X. (2008) A family of isolated interleaved boost and buck converters with winding-cross-coupled inductors. *IEEE Transactions on Power Electronics*, **23** (6), 3164–3173.

13. Raggl, K., Nussbaumer, T., Doerig, G. *et al.* (2009) Comprehensive design and optimization of a high-power-density single-phase boost PFC. *IEEE Transactions on Industrial Electronics*, **56** (7), 2574–2587.

14. Nussbaumer, T., Raggl, K. and Kolar, J.W. (2009) Design guidelines for interleaved single-phase boost PFC circuits. *IEEE Transactions on Industrial Electronics*, **56** (7), 2559–2573.

15. Cervantes, I., Mendoza-Torres, A., Garcia-Cuevas, A.R., and Perez-Pinal, F.J. (2009) Switched control of interleaved converters. Proceedings of the IEEE VPPC, Sepember 7–11, pp. 1156–1161.

16. Yang, B., Li, W., Zhao, Y. and He, X. (2010) Design and analysis of a grid-connected photovoltaic power system. *IEEE Transactions on Power Electronics*, **25** (4), 992–1000.

17. Ku, C.P., Chen, D., Huang, C.S. and Liu, C.Y. (2011) A Novel SFVM-M3 control scheme for interleaved CCM/DCM boundary-mode boost converter in PFC applications. *IEEE Transactions on Power Electronics*, **26** (8), 2295–2303.

18. Jung, D.Y., Ji, Y.H., Park, S.H. *et al.* (2011) Interleaved soft-switching boost converter for photovoltaic power-generation system. *IEEE Transactions on Power Electronics*, **26** (4), 1137–1145.

19. Kim, H., Falahi, M., Jahns, T.M. and Degner, M.W. (2011) Inductor current measurement and regulation using a single DC link current sensor for interleaved DC–DC converters. *IEEE Transactions on Power Electronics*, **26** (5), 1503–1510.

20. Asiminoaei, L., Aeloiza, E., Enjeti, P.N. and Blaabjerg, F. (2008) Shunt active-power-filter topology based on parallel interleaved inverters. *IEEE Transactions on Industrial Electronics*, **55** (3), 1175–1189.

21. Zhang, D., Wang, F., Burgos, R. *et al.* (2010) Impact of interleaving on AC passive components of paralleled three-phase voltage-source converters. *IEEE Transactions on Industry Applications*, **46** (3), 1042–1054.

22. Singh, B.N., Joos, G., and Jain, P. (2000) A new topology of 3-phase PWM AC/DC interleaved converter for telecommunication supply systems. IEEE Industry Applications Conference, Vol. 4, pp. 2290–2296.

23. Singh, B.N., Joos, G., and Jain, P. (2000) A new topology of 3-phase PWM AC/DC interleaved converter for telecommunication supply systems. IEEE Industry Applications Conference, October 8–12, Vol. 4, pp. 2290–2296.

24. Tamyurek, B. and Torrey, D.A. (2011) A three-phase unity power factor single-stage AC–DC converter based on an interleaved flyback topology. *IEEE Transactions on Power Electronics*, **26** (1), 308–318.

25. García, O., Zumel, P., de Castro, A. and Cobos, J.A. (2006) Automotive DC–DC bidirectional converter made with many interleaved buck stages. *IEEE Transactions on Power Electronics*, **21** (3), 578–586.

26. Gerber, M., Ferreira, J.A., Hofsajer, I.W., and Seliger, N. (2004) Interleaving optimization in synchronous rectified DC/DC converters, interleaving optimization in synchronous rectified DC/DC converters. Proceedings of the 35th Annual IEEE PESC, June 20-25, Vol. 6, pp. 4655–4666.

27. Oliver, J.A., Zumel, P., García, O. *et al.* (2004) Passive component analysis in interleaved buck converters. Proceedings of the IEEE Applied Power Electronics Conference (APEC'04), Vol. 1, pp. 623–628.

28. Mohan, N., Undeland, T.M. and Robbins, W.P. (2003) *Power Electronics, Converters, Applications, and Design*, John Wiley & Sons, Inc., Hoboken, NJ.

29. Steinmetz, C.P. (1984) On the law of hysteresis. *Proceedings of the IEEE*, **72**, 196–221.

30. Liserre, M., Blaabjerg, F. and Hansen, S. (2005) Design and control of an LCL-filter-based three-phase active rectifier. *IEEE Transactions on Industry Applications*, **41** (5), 1281–1291.

31. Epcos http://www.epcos.com, (accessed January 2012).

32. IEEE (1992) IEEE Std 519. Recommended Practices and Requirements for Harmonic Control in Electrical Power Systems, IEEE.

33. Ogata, K. (1995) *Discrete-Time Control Systems*, Prentice Hall.

34. Ito, Y. and Kawauchi, S. (1995) Microprocessor-based robust digital control for UPS with three-phase PWM inverter. *IEEE Transactions on Power Electronics*, **10** (2), 196–203.

35. Moreno, J.C., Huerta, J.M.E., Gil, R.G. and Gonzalez, S.A. (2009) A Robust predictive current control for three-phase grid-connected inverters. *IEEE Transactions on Industrial Electronics*, **56** (6), 1993–2004.

36. Espi, J.M., Castello, J., Fischer, J.R. and Garcia-Gil, R. (2010) A synchronous reference frame - Robust predictive current control for three-phase grid-connected inverters. *IEEE Transactions on Industrial Electronics*, **57** (3), 954–962.

8

Repetitive Current Control of an Interleaved Grid-Connected Inverter

8.1 Introduction

Different controllers and topologies [1–4] have been used for grid-connected inverters to obtain high quality output current. However, classical PID controllers and their derivatives suffer from relatively low loop gain at the fundamental frequency and its harmonics and hence can have poor grid harmonic disturbance rejection, which results in poor output current THD (total harmonic distortion) if the grid voltage THD is relatively high. Proportional-resonant (PR) controllers have been widely used for grid-connected inverters due to their ability to reject individual harmonics [5–9]. Theoretically, this introduces an infinite gain at a selected frequency. By having multiple PR controllers, higher order harmonics can be eliminated but a bank of resonant controllers increases the complexity of the system and the calculation burden of the digital signal processor (DSP).

Repetitive feedback control (RC), which is based on the concept of iterative learning control, has been widely used for many practical industrial systems, such as manufacturing [10] and robotics [11]. In these controllers error between the reference and the output over one fundamental cycle is used to generate a new reference for the next fundamental cycle. RC is mathematically equivalent to a parallel combination of an integral controller, an infinite number of resonant controllers, and a proportional controller [12]. It has the advantage of being simpler to implement than PR controllers. However, it creates resonance gain peaks at high frequencies, which can lead to instability. A low-pass filter can be used to attenuate the high frequency resonance gain peaks but if the bandwidth of the plant is low (i.e., the gain crossover frequency is not very high with respect to the fundamental and low harmonic frequencies) such a filter will reduce the low frequency resonance gains and, consequently, will deteriorate the performance of the RC [13].

Power Electronic Converters for Microgrids, First Edition. Suleiman M. Sharkh, Mohammad A. Abusara, Georgios I. Orfanoudakis and Babar Hussain.
© 2014 John Wiley & Sons, Ltd. Published 2014 by John Wiley & Sons, Ltd.
Companion Website: www.wiley.com/go/sharkh

The application of RC to the control of grid-connected and stand-alone inverters is an active area of research [14–16]. A number of papers describe the effect of parameter variations on overall system stability and transient response. However, limits of performance of RC with reference to system bandwidth need to be investigated, especially for two-level grid-connected inverters.

The high bandwidth requirement of the inverter and its output filter dictates a high PWM (pulse width modulation) switching frequency, which becomes challenging in high power systems as the maximum achievable switching frequency of power electronic devices reduces as their power rating increases. This limitation can be overcome by using an interleaved inverter topology, such as that reported in [17] and Chapter 7. Interleaving is a form of paralleling technique where a single converter channel, for example, half a bridge with an output inductor, is replaced by N smaller channels connected in parallel, as shown in Figure 8.1, with their switching instants phase shifted equally over the switching period.

By introducing a phase shift between the switching instants of the parallel channels of each phase, the amplitude of the total ripple current is N times less and its frequency is N times greater than that of a conventional inverter. The reduction in total current ripple amplitude and the increase in its frequency reduce the size of the required filter capacitance considerably, as mentioned earlier. Furthermore, sharing the current among a number of channels enables the use of smaller lower current power switches which can switch at high frequency, thus allowing a reduction in inductor size. The net result is that the size of the filter and the overall size of the system are smaller than an equivalent classical two-level bridge inverter with LCL output filter. The reduction in total current ripple and the high frequency also eliminate the need for a second output filter inductor (i.e., using an LCL filter instead of an *LC* filter) that is sometimes used in two-level and multi-level grid-connected inverters to block the switching ripple in cases where the grid impedance is too low. Consequently, the bandwidth of the interleaved system is high due to the much smaller output filter capacitors and equivalent inductor. This makes RC, which requires a high system bandwidth, an attractive option for the control of interleaved inverters.

In this chapter the design and practical implementation of a repetitive controller for a six-channel interleaved inverter are discussed. Simulation and experimental results are also presented to demonstrate the effectiveness of the proposed controller in improving the THD of the output current of the inverter.

8.2 Proposed Controller and System Modeling

Figure 8.2 shows the block diagram of the PWM inverter, LC filter, and the control system of one of the phases. This controller structure uses only one current sensor of the total inductors' current as opposed to the controller structure presented in chapter 7. However, for a protection reason, the individual inductor currents are still measured and monitored using six small current sensors. Active damping was not implemented as the sampling frequency f_{sw} (35 kHz) is quite low compared to the filter resonance frequency. Instead, a resistor in series with the filter capacitor is

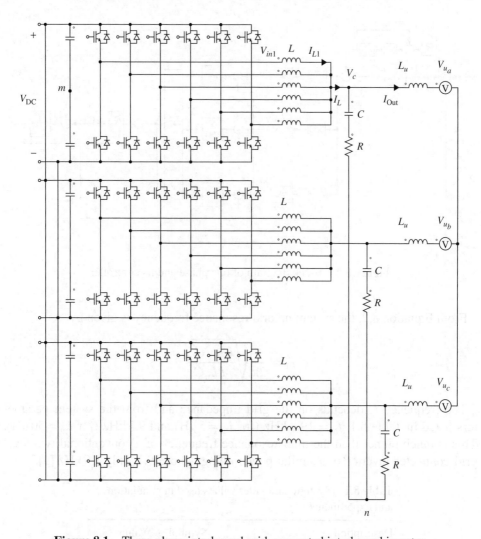

Figure 8.1   Three-phase interleaved grid-connected interleaved inverter

used to provide passive damping. Fortunately, due to the ripple cancellation feature of the interleaved topology, the capacitor current is quite small and hence the power dissipation in R is negligible as discussed in chapter 7.

From Figure 8.2, the total inductor current I_L can be shown to be given by

$$I_L = (V_{inx} - B(s)V_u)A(s) \quad (x = 1, 2, .., N) \tag{8.1}$$

where

$$A(s) = \frac{N(L_u Cs^2 + RCs + 1)}{LL_u Cs^3 + RC(L + NL_u)s^2 + (L + NL_u)s} \tag{8.2}$$

$$B(s) = \frac{RCs + 1}{L_u Cs^2 + RCs + 1} \tag{8.3}$$

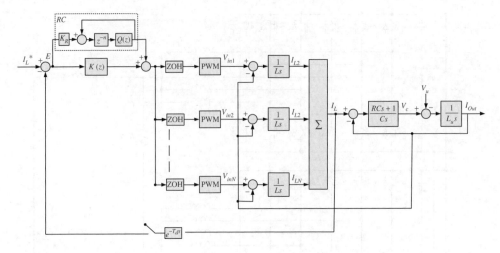

Figure 8.2   Block diagram of one phase and its controller

From Equation 8.2, the system natural resonance frequency is given by

$$f_n = \frac{1}{2\pi}\sqrt{\frac{L + NL_u}{LL_u C}} \tag{8.4}$$

The frequency f_n depends on the grid impedance and from the system parameters listed in Table 8.1, $f_n = 19.8\,\text{kHz}$ (for $L_u = 5\,\mu\text{H}$) and $9.3\,\text{kHz}$ (for $L_u = 50\,\mu\text{H}$). This is much higher than the filter resonance frequency of a conventional two-level grid-connected inverter, of a similar power rating, whose f_n is only $1.5\,\text{kHz}$ [3].

Table 8.1   System parameter values used in simulation and experiments

Description	Symbol	Value
Number of interleaved channels	N	6
Passive damping resistor	R	$0.5\,\Omega$
Channel inductor	L	$190\,\mu\text{H}$
Filter capacitor	C	$15\,\mu\text{F}$
Switching frequency	f_{sw}	$35\,\text{kHz}$
Sampling frequency	f_s	$35\,\text{kHz}$
Number of samples in one cycle	n	700
Time delay	T_d	$14.28\,\mu\text{s}$
Grid voltage	V_u	230 V (rms)
Grid frequency	f_o	50 Hz
Inverter DC voltage	V_{DC}	700 V DC
Inverter rated current	I_{Out}	90 A (rms)

Equations 8.2 and 8.3 are represented by the block diagram in Figure 8.3. In the discrete time domain, the open loop transfer function, including the DSP computational time delay T_d and zero order hold, is obtained by performing the z-transform,

$$G(z) = Z\left[\frac{1 - e^{-T_s s}}{s} e^{-T_d s} A(s)\right] \tag{8.5}$$

where T_s is the sampling period. The repetitive current controller transfer function is given by

$$G_{RC}(z) = \frac{K_R Q(z) z^{-n}}{1 - Q(z) z^{-n}} \tag{8.6}$$

where, $Q(z)$ is a low pass filter, K_R is the repetitive controller gain, and n is the number of samples in one fundamental cycle. A simplified block diagram of the system is shown in Figure 8.4 where D_u represents the grid disturbance $B(s)V_u$.

8.3 System Analysis and Controller Design

The controller design involves the determination of $K(z)$, $Q(z)$, and K_R. The controller gain $K(z)$ can be a phase lag to improve harmonics rejection, as was discussed in

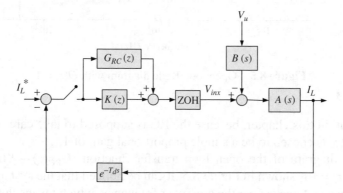

Figure 8.3 Single channel equivalent block diagram

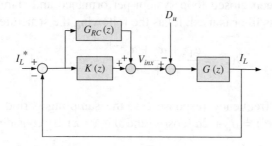

Figure 8.4 Simplified block diagram

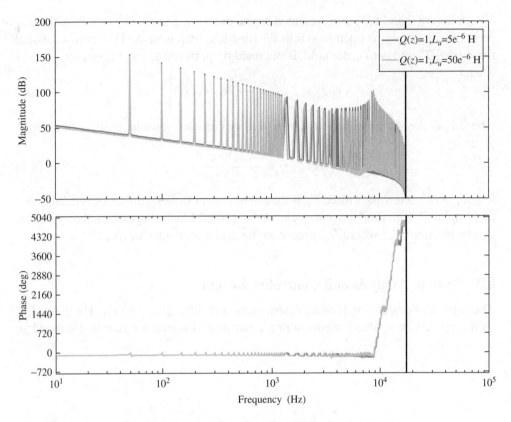

Figure 8.5   Open loop Bode diagram with $Q(z) = 1$

chapter 7 but, in this chapter, because the RC is supposed to take care of harmonics rejection, $K(z)$ is chosen to be a simple proportional gain of 1.

The Bode diagram of the open loop transfer function $(G_{RC}(z) + K(z))G(z)$ with $K(z) = 1$, $Q(z) = 1$ is shown in Figure 8.5. It can be noticed that the system is unstable due to the resonant peaks near the crossover frequency, which means that $Q(z)$ needs to be modified to attenuate the high frequency peaks. The selection of $Q(z)$ reflects the trade-off between closed loop system performance and stability robustness. A zero-phase low pass filter is used; it has the following the structure:

$$Q(z) = \frac{\alpha_1 z + \alpha_o + \alpha_1 z^{-1}}{\alpha_o + 2\alpha_1}, \quad \alpha_o, \alpha_1 < 1 \tag{8.7}$$

The normalized frequency response (i.e. the sampling period T_s is set to 1) of Equation 8.7 is $Q(e^{j\omega}) = \alpha_o + 2\alpha_1 \cos(\omega)$ and $\omega \in (0, \pi)$. By considering $\alpha_o + 2\alpha_1 = 1$, for unit gain response, the magnitude of $Q(j\omega)$ can be written as follows:

$$|Q(j\omega)| = \begin{cases} \alpha_o + 2\alpha_1 & \omega = 0 \\ \alpha_o + 2\alpha_1 \cos(\omega) & \omega(0,\pi) \\ \alpha_o - 2\alpha_1 & \omega = \pi \end{cases} \quad (8.8)$$

By selecting $\alpha_o = 0.5$ and $\alpha_1 = 0.25$, then

$$|Q(j\omega)| = \begin{cases} 1 & \omega = 0 \\ 0 < 0.5\,(1 + \cos\omega) < 1 & \omega \in (0,\pi) \\ 0 & \omega = \pi \end{cases} \quad (8.9)$$

Equation 8.9 concludes that $|Q(j\omega)|$ is 1 at low frequencies and 0 at high frequencies. The filter is now given by

$$Q(z) = 0.25z + 0.5 + 0.25z^{-1}. \quad (8.10)$$

The Bode diagram of $Q(z)$ as described in Equation 8.10 is shown in Figure 8.6. The value of the RC gain K_R needs to be carefully selected as it is a key parameter for error convergence and system stability. A high repetitive controller gain K_R results in

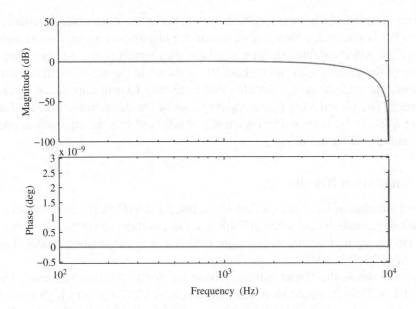

Figure 8.6 Bode diagram of $Q(z) = 0.25z + 0.5 + 0.25z^{-1}$

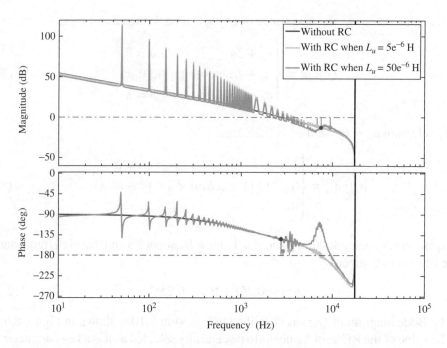

Figure 8.7 Bode diagram of system with $Q(z) = 0.25z + 0.5 + 0.25z^{-1}$, $K_R = 0.1$

fast error convergence but the feedback system becomes less stable. A suitable value of $K_R = 0.1$ is selected, which gives a satisfactory transient response while maintaining stability. After suitable selection of all the above parameters, the open loop Bode diagram of the system with and without RC is shown in Figure 8.7 with two different values of grid impedance L_u. Stability and immunity to grid impedance variation is confirmed. The closed loop frequency response of the disturbance transfer function relating $I_L(z)/D_u(z)$ is shown in Figure 8.8 which confirms the attenuation provided by RC at the grid harmonics.

8.4 Simulation Results

Detailed simulation has been carried out using the MATLAB® SimPowerSystems. The system parameters are listed in Table 8.1. Grid voltage harmonics were measured in the laboratory and similar values were included in the simulation model. The total grid voltage THD is 1.9%.

Figure 8.9 shows the output current without RC for a 10 A (rms) demand. The current THD is 22%. It should be noted that the current THD is very high because the current demand is only 11% of the rated current of the inverter and the controller proportional gain $K(z)$ is set to 1. The current THD could also be improved if $K(z)$ was set to a higher value or modified to be a phase lag. Figure 8.10 shows the steady-state output current (the time scale has been adjusted and the starting time shown in the figure is after the controller has reached steady state). The current THD is reduced

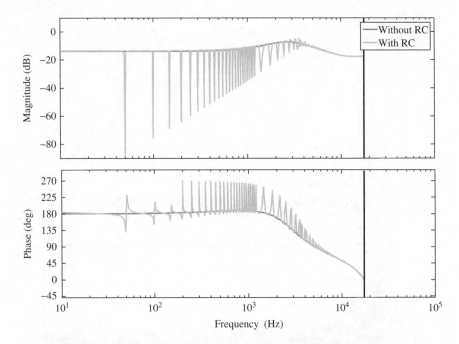

Figure 8.8   Frequency response of the disturbance transfer function $\frac{I_L(z)}{D_u(z)}$

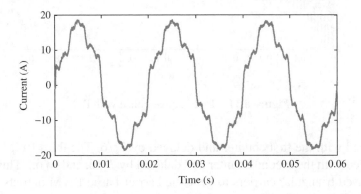

Figure 8.9   Simulated output current without RC

to only 2.2%. The effectiveness of the RC in improving the current THD is clearly demonstrated. Figure 8.11 shows the current error and it can be noticed that the error converges within about 0.25 s.

8.5   Experimental Results

The proposed controller was tested experimentally with an interleaved inverter. The system parameters are listed in Table 8.1. The controller was implemented using the Texas Instrument TMS320F2808 32-bit DSP. This processor has the capability of generating six interleaved PWM outputs. The internal counter of the first PWM carrier is set to give the required switching frequency. The second PWM counter is

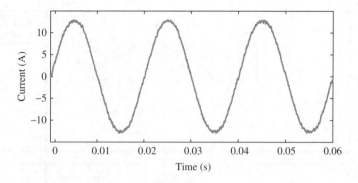

Figure 8.10   Simulated output current with RC

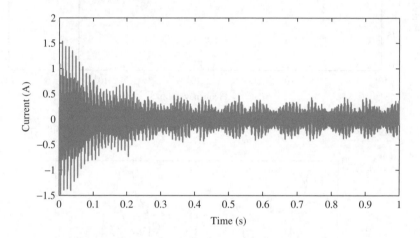

Figure 8.11   Error convergence with RC

synchronized with the first counter and delayed by $T_s/6$. The third PWM counter is synchronized with the second counter and delayed by $T_s/6$, and so on. This produces six interleaved triangular carriers to generate carrier-based PWM outputs. One DSP per phase was used and the low speed communications between the controllers, such as start/stop and total current commands, were implemented using the controller area network (CAN) protocol. Synchronization with the grid was implemented by having each phase controller measuring its corresponding grid phase voltage to detect the zero crossing. The reference sine waves were generated internally by the individual controllers using look-up tables of 700 samples (35 kHz/50 Hz). The sine wave amplitude is set externally (by a setting in the user interface) and sent via CAN-bus to the three phase controllers. The input DC is regulated by an external boost circuit to 700 V DC.

The RC was implemented indirectly by creating an array of 700 entries representing a state variable x which is the difference between RC input and output such as $x(i) = K_R e(i) + y(i)$, as illustrated in Figure 8.12.

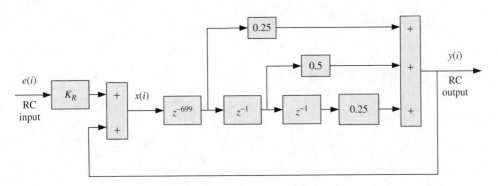

Figure 8.12   RC implementation

Figure 8.13 shows the output current when RC is de-activated. The demand current is set to 15 A (rms). The current THD is measured to be 16.0%. Figure 8.14 shows the output current but when the RC is activated. The current THD is measured to be only 2.0%. In the system reported in [17] (which is similar to this one but without RC and

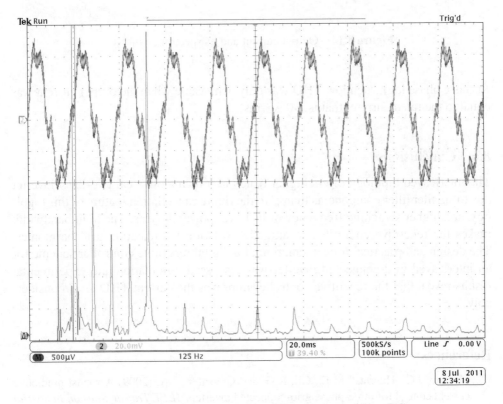

Figure 8.13   Output current and its spectrum without RC

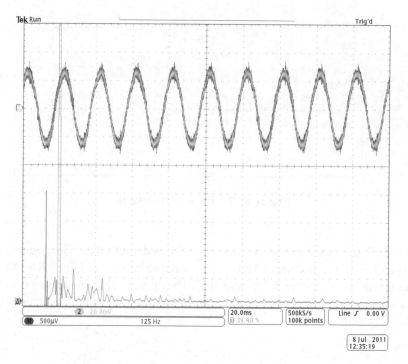

Figure 8.14   Output current and its spectrum with RC

$K(z)$ is a phase lag), such low THD was only achievable when the output current was equal to the rated current, that is, 90 A (rms).

8.6   Conclusions

The interleaved topology offers higher bandwidth than a classical two-level inverter due to smaller filter components thanks to the ripple cancellation feature of this topology, and higher switching frequency of its lower power devices. The high bandwidth makes the repetitive controller an attractive solution for rejecting grid harmonics. The design and practical implementation of a digital repetitive controller scheme for an interleaved grid-connected inverter was discussed. Simulation and experimental results reveal that the repetitive controller improves the current THD level considerably.

References

1. Moreno, J.C., Huerta, J.M.E., Gil, R.G. and Gonzalez, S.A. (2009) A robust predictive current control for three-phase grid-connected inverters. *IEEE Transactions on Industrial Electronics*, **56** (6), 1993–2004.

2. Zhao, W., Lu, D.D.C. and Agelidis, V.G. (2011) current control of grid-connected boost inverter with zero steady-state error. *IEEE Transactions on Power Electronics*, **26** (10), 2825–2834.
3. Abusara, M.A. and Sharkh, S.M. (2011) Digital control of a three-phase grid connected inverter. *International Journal on Power Electronics*, **3** (3), 299–319.
4. Shuitao, Y., Qin, L., Peng, F.Z. and Zhaoming, Q. (2011) A robust control scheme for grid-connected voltage-source inverters. *IEEE Transactions on Industrial Electronics*, **58** (1), 202–212.
5. Teodorescu, R., Blaabjerg, F., Liserre, M. and Loh, P.C. (2006) Proportional-resonant controllers and filters for utility-connected voltage-source converters. *IEE Proceedings on Electric Power Applications*, **153**, 750–762.
6. Hwang, J.G., Lehn, P.W. and Winkelnkemper, M. (2010) A generalized class of stationary frame-current controllers for utility-connected AC-DC converters. *IEEE Transactions on Power Delivery*, **25**, 2742–2751.
7. Eren, S., Bakhshai, A., and Jain, P. (2011) Control of three-phase voltage source inverter for renewable energy applications, IEEE International Telecommunications Energy Conference (INTELEC), Amsterdam, The Netherlands.
8. Shen, G., Zhu, X., Zhang, J. and Xu, D. (2010) A new feedback method for PR current control of LCL-filter based utility-connected inverter. *IEEE Transactions on Industrial Electronics*, **57** (6), 2033–2041.
9. Liserre, M., Teodorescu, R. and Blaabjerg, F. (2006) Multiple harmonics control for three phase utility converter systems with the use of PI-RES current controller in a rotating frame. *IEEE Transactions on Power Electronics*, **21** (3), 836–841.
10. Chen, S.L. and Hsieh, T.H. (2007) Repetitive control design and implementation for linear motor machine tool. *International Journal of Machine Tools and Manufacture*, **47**, 1807–1816.
11. Tinone, H. and Aoshima, N. (1996) Parameter identification of robot arm with repetitive control. *International Journal of Control*, **63**, 225–238.
12. Lu, W., Zhou, K., and Yang, Y. (2010) A general internal model principle based control scheme for CVCF PWM converters. 2nd IEEE International Symposium on Power Electronics for Distributed Generation Systems (PEDG), Hefei, China.
13. Jamil, M. (2012) Repetitive current control of two-level and interleaved three-phase PWM utility connected converters. PhD Thesis, University of Southampton.
14. Gao, J., Zheng, Q.T. and Lin, F. (2011) Improved deadbeat current controller with a repetitive -control-based observer for PWM rectifiers. *Journal of Power Electronics*, **11**, 64–73.
15. Hornik, T. and Zhong, Q.C. (2011) A current-control strategy for voltage-source inverters in microgrid based on H-infinity and repetitive control. *IEEE Transactions on Power Electronics*, **26**, 943–952.
16. Jiang, S., Cao, D., Li, Y. *et al.* (2012) Low-THD, Fast-transient, and cost-effective synchronous-frame repetitive controller for three-phase UPS inverters. *IEEE Transactions on Power Electronics*, **27**, 2994–3005.
17. Abusara, M.A. and Sharkh, S.M. (2010) Design of a robust digital current controller for a grid connected interleaved inverter. IEEE International Symposium on Industrial Electronics (ISIE), pp. 2903–2908.

9

Line Interactive UPS

9.1 Introduction

To increase the reliability of a microgrid, energy storage systems are essential [1]. Energy is stored while in grid-connected mode, when the microgrid's DG (distributed generator) systems produce excess power, to be used later to supply critical loads during power outages. In stand-alone mode, they can be used to boost the power supplied by the microgrid if the DG systems cannot meet the expected level of power. To meet these demands, the energy storage system needs to be able to work in grid-connected and stand-alone modes. In the latter mode of operation, the system needs to operate in parallel with other DG systems to meet the variable power demand of the load. More importantly, it needs to switch seamlessly between the two modes. Line interactive UPS (uninterruptible power supply) systems are good candidates for providing energy storage within microgrids as they can be connected in parallel with both the main grid and the local load.

In the classical topology of line interactive UPS systems [2–7], the battery is interfaced with a bidirectional DC/AC inverter which is connected in parallel with the critical load. The load and the inverter are connected to the grid via an inductor. The inverter charges the battery during normal operation and supplies the critical load during grid power outage. Sometimes a dedicated DC/DC converter is used between the battery and the DC/AC inverter. Compared with on-line double conversion UPS, the line interactive UPS is simpler, cheaper, more efficient, and more reliable. This topology, however, does not provide voltage regulation to the load. An alternative topology of the line interactive UPS – known as series-parallel or delta conversion line-interactive UPS – incorporates a complementary DC/AC inverter connected between the battery and the AC mains via a series transformer [8–14]. In this configuration, voltage regulation of the load becomes possible but at the expense of lower efficiency and extra size and cost due to the use of an extra inverter and bulky transformer. However, this topology is still more efficient than classical on-line

Power Electronic Converters for Microgrids, First Edition. Suleiman M. Sharkh, Mohammad A. Abusara, Georgios I. Orfanoudakis and Babar Hussain.
© 2014 John Wiley & Sons, Ltd. Published 2014 by John Wiley & Sons, Ltd.
Companion Website: www.wiley.com/go/sharkh

double conversion UPS because the complementary inverter has only to supply 10–20% of the UPS nominal power [15].

There are a number of publications on the control of line interactive UPS systems [16–22]. Tirumala *et al.* [16] proposed a control algorithm for grid interactive PWM (pulse width modulation) inverters to obtain a seamless transfer from grid-connected mode to stand-alone mode and vice versa. When it is connected to the grid, the inverter operates in current-controlled mode, regulating the current injected into the point of common coupling (PCC). In stand-alone mode, however, the inverter operates in voltage-controlled mode, regulating the output voltage across the load. When the grid fails, the controller transfers from current-controlled mode to voltage-controlled mode with the voltage reference set to the value measured at the PCC. When the grid fault is cleared, the inverter voltage is adjusted such that its amplitude and phase match that of the grid. The isolating switch (that connects the inverter to the PCC) is then closed and the controller transfers from voltage-controlled mode to current-controlled mode, with the current reference set initially to the load current (measured before closing the isolating switch). The reference current is then ramped up or down to the required value. Similar approaches were also reported in [17, 18]. The main disadvantage of these systems is that the load voltage after the grid fails and before starting the transitional period (from current-controlled to voltage-controlled mode) depends on the current demand and the load. This might result in a very high or very low voltage across the load during the transitional period. A notch filler that is used to mitigate these voltage transients was proposed in [19]. Kim *et al.* [20] proposed a controller for truly seamless transfer from grid-connected mode to stand-alone mode. The controller has an inner voltage control loop that regulates the output voltage. During grid-connection, the power angle (the angle between the inverter voltage and the voltage at the PCC) of the output voltage is set by an outer control loop, depending on the required injected grid current. When the grid fails, the power angle is saturated to a maximum pre-defined value. In the system reported in [21], the power angle between the UPS voltage and the grid voltage is measured by a phase detector and adjusted to control the power. The line interactive UPS system proposed in [22] has the voltage controller and the current controller working in parallel in order to achieve seamless transfer between the two modes.

Although the controllers reported above enable seamless transfer of a single unit from grid-connected and stand-alone mode, the main disadvantage is that the UPS units are not capable of operating autonomously in parallel with other DG units and thus they cannot form a microgrid.

Chandorkar *et al.* [23] proposed a line interactive UPS system based on $P–\omega$ and $Q–V$ droop control, where the inverter frequency and voltage amplitude are drooped linearly with the inverter output active and reactive power, respectively. UPS units can operate in parallel and load sharing is achieved without the need for communication signals between the inverters. Thus, a flexible distribution architecture is achieved and UPS units can operate in a microgrid. In this proposed configuration, each UPS unit consists of a battery, DC/AC inverter, and LC filter. The UPS units are connected to a common bus via inductors. Each UPS unit has its own local load that is connected across the filter capacitor. This configuration provides immunity from disturbances on the secure bus, and has reduced filter interactions, but this is obtained at the expense of system flexibility. In [24], a line interactive UPS system that is capable of operating in a microgrid was proposed. The controller of the UPS inverter is based on two control loops: an inner voltage feedback loop that regulates the output voltage and an outer active and reactive power sharing loop which is implemented by drooping control. Both systems in [23, 24] provide truly seamless transfer between grid-connected mode and stand-alone parallel mode and vice versa. However, the integration of the battery and its DC/DC converter into the system was not studied. Also, the stability of the DC-link voltage was not discussed. This becomes an important issue, taking into account the requirement to transfer seamlessly from battery charging mode in grid-connected mode to battery discharging in stand-alone paralleling mode.

This chapter presents and discusses a line interactive UPS system to be used within a microgrid, as illustrated in Figure 9.1. The system can transfer practically seamlessly between grid-connected and stand-alone modes, sharing the load within a microgrid in parallel with other DG units. The control of the complete system including the battery and its bidirectional DC/DC converter is considered. Seamless transition between grid-connected and stand-alone parallel modes is made possible by using a frequency and voltage drooping method to control active and reactive power flow.

The use of frequency and voltage drooping methods to control power sharing of parallel-connected PWM inverters is well established in the literature [23–38]. This method requires active and reactive power to be measured in order to droop the frequency and voltage accordingly. In reported systems [24–34], a low pass filter is used to obtain the average power from the single phase instantaneous power. In a balanced

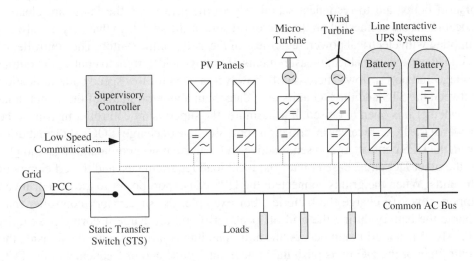

Figure 9.1   Microgrid structure

three-phase system, the instantaneous power equals the average power and such a filter might not be necessary. However, in an unbalanced three-phase system, the instantaneous power has a ripple component and a filter becomes essential to prevent the ripple component from propagating to the frequency and amplitude as a result of drooping control. This filter has a significant effect on the dynamics of drooping control due to the associated phase lag [34]. In [35], the average power was measured by integrating the instantaneous power, however, no discussion regarding the advantage of this method over the low pass filter was provided. In this chapter it is shown that the real time integration method for calculating average power gives superior controller dynamic performance compared to the low pass filter method.

In the proposed controller, the current injected into the grid is basically controlled by adjusting the power angle and, hence, there is no direct control over the quality of the current, in contrast to traditional current mode grid-connected inverters [39−41]. Therefore, in the proposed controller virtual impedances are utilized as an effective method for reducing the effects of grid voltage harmonics on current quality.

The chapter is organized as follow. Section 9.2 gives an overview of the system. Section 9.3 discusses the core controller of the DC/AC converter. Section 9.4 discusses the proposed power flow controller. In Section 9.5, the design of a DC-link voltage controller that sets the demand for the power flow controller is discussed. Finally, experimental results are presented in Section 9.6.

9.2 System Overview

The general overall structure of a microgrid was shown in Figure 9.1. It consists of DG units, UPS units, local loads, supervisory controller, and a static transfer switch (STS). The STS is used at the PCC to isolate the microgrid from the grid in the case of grid faults and to reconnect seamlessly to the grid when the faults are cleared. Local loads are connected on the microgrid side of the switch so that they are always supplied with electrical power, regardless of the status of the switch. The controller of each DG system uses local measurements of voltage and current to control the output voltage and power flow. There is still, however, a need for low speed communication between the STS and the DG units to update them about the status of the switch, that is, whether it is open or closed. Furthermore, the supervisory controller measures the power at the PCC, receives information about the state of charge (SOC) of the batteries from the UPS controllers and sends active and reactive power commands P_{ref} and Q_{ref} to the other DG units, based on tariffs, local load requirements, and the efficiency of the units. When the grid is connected, the UPS can export power to the grid or it can import power to charge the batteries. For example, the supervisory controller may charge the battery during the grid off-peak tariff and discharge it during peak tariff periods. When a grid fault occurs, the anti-islanding controllers embedded inside the controllers of the DG units push the voltage amplitude and/or frequency at the PCC toward their upper or lower limits. Once the amplitude or frequency upper or lower limits are reached, the controller of the STS decides that the grid is not healthy and opens the switch. It also sends a signal to the DG units to update them about the status of the switch. The DG units carry on supplying power in parallel according to their ratings, sharing the power demanded by the distributed local loads. After the grid fault is cleared, the STS reconnects the microgrid; it monitors the voltage signals on both sides and closes when these signals are in phase to ensure transient free operation. It also sends an update signal to the DG units within a short period (~ 20 ms) after closing the switch.

It is important to note that using the above anti-islanding strategy the time period from grid failure until the opening of the STS may vary depending on the mismatch between the power generated by the microgrid and the local distributed load. If the mismatch is large, the voltage and frequency shift toward their upper or lower limits quite quickly. However, if the mismatch is small, it will take longer for the anti-islanding controller to cause the voltage/frequency to shift toward the limits. In the worst case scenario of perfect mismatch between the generated power and the load, the anti-islanding controller should not take more than 2 s according to the IEEE Standard 1547 [42]. This time is not critical for the operation of the microgrid, which is able to carry on working thanks to its autonomous feature. Once islanding is detected by the STS controller, the switch is opened and a signal is sent to the DG units so that the drooping-based power flow controller is reconfigured. This involves modifying the drooping function slightly (as will be explained in Section 9.4) so that DG units within the microgrid share the local load equitably. It is also worth mentioning that the STS may take a few milliseconds to open. However, this time is also not critical because the local load has a continuous supply of power during this transitional period.

The microgrid considered in this chapter consists of two 60 kW line interactive UPS systems, each powered by a 30 Ah (550–700 V) lithium-ion battery. The circuit diagram of each UPS system is shown in Figure 9.2. The UPS system consists of a battery, a bidirectional DC/DC converter, and a bidirectional three-phase voltage source inverter/active rectifier with an output LCL filter. The DC/AC converter parameters are given in Table 9.1. The UPS proposed controller is shown in Figure 9.3. Figure 9.3a shows the controller of the bidirectional DC/AC converter. An outer power

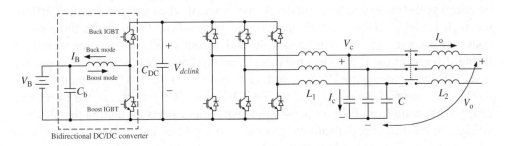

Figure 9.2   Circuit diagram of the UPS system

Table 9.1 DC/AC converter parameter values

Symbol	Value	Description
L_1	350 µH	Inverter-side filter inductor
C	160 µF	Filter capacitor
L_2	250 µH	Grid-side filter inductor
C_{DC}	2000 µF	DC link capacitor
P_{max}	60 kW	Maximum active power rating
Q_{max}	45 kVAR	Maximum reactive power rating
V_{DClink_max}	1000 V	Maximum allowed transient value of the DC-link voltage (trip limit)

flow controller sets the voltage demand for an inner voltage core controller loop. The design of these controllers will be discussed in the following sections. When the power flows from the grid to the battery, the DC/DC converter operates in buck mode and the boost IGBT is held open. When the power flows in the opposite direction, the buck IGBT is held open and the DC/DC converter operates in boost mode, regulating the DC-link voltage to a suitable level in order to inject power into the grid.

Figure 9.3b shows the controller of the bidirectional DC/DC converter. During battery charging mode, the buck IGBT is modulated and, depending on the battery voltage, the controller operates either in current mode or voltage mode regulating the battery current or voltage, respectively. When the battery discharges, the boost IGBT is modulated to regulate the DC-link voltage. During battery charging, the DC-link voltage is controlled by the three-phase converter, which in this case operates as an active rectifier. When discharging, however, the DC-link voltage is controlled by the DC/DC converter, operating in boost mode, as mentioned earlier. To decouple these two controllers, which control a common DC-link voltage, the voltage demand for the active rectifier $V^*_{dclink\_2}$ is set to be higher (800 V) than the voltage demand of the boost $V^*_{dclink\_1}$ (750 V). To understand how the controllers react during a sudden transition between operating modes, the following scenario is considered: Suppose that the UPS is in charging mode, the DC-link in this case will be regulated by the three-phase converter operating as an active rectifier and the DC/DC converter will be charging the battery in buck mode. At the moment when the grid fails and before the fault is detected by the STS switch, the power flow through the three-phase converter changes direction immediately and automatically as a consequence of losing stiffness at the PCC. Thus the power starts to flow from the DC-link capacitor to the AC load causing the DC-link voltage to drop from the $V^*_{dclink\_2}$ demand. Once, the DC-link drops below $V^*_{dclink\_1}$, the DC/DC controller recognizes the event and changes its mode from buck to boost immediately and starts regulating the DC-link voltage. In this scheme, the smooth transition from grid-connected charging mode to stand-alone mode is possible without relying on external communication. When the grid fault is cleared, the STS closes and sends an update signal to the UPS units.

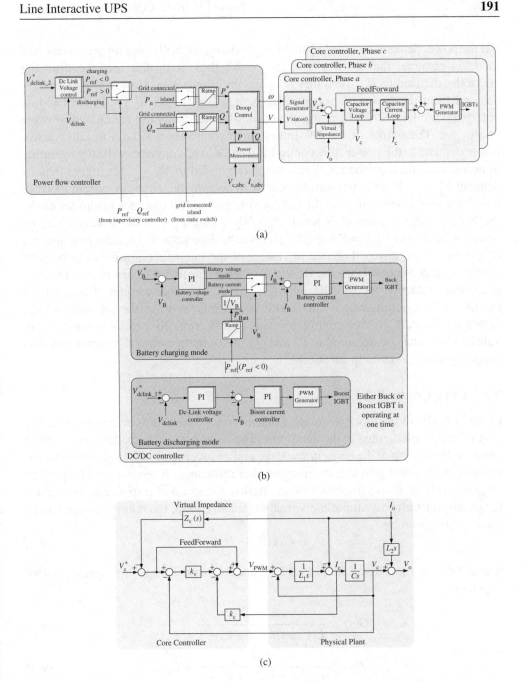

Figure 9.3 Line interactive UPS controller. (a) DC/AC (or AC/DC) controller, (b) DC/DC controller, and (c) transfer function block diagram for one phase and its core controller

If the power demand received from the supervisory controller during grid-connected mode is positive, that is, $P_{ref} > 0$, the power flow controller sets the drooping controller demand to P_{ref} such that $P^* = P_{ref}$ in Figure 9.3a. However, if the power demand from the supervisory controller is negative, that is, $P_{ref} < 0$, the DC-link voltage controller, within the power flow controller, starts to raise the DC-link voltage to $V^*_{dclink\_2}$. The output of the DC-link voltage controller will be a negative active power demand to the power flow controller. The DC/DC converter will stop operating in boost mode because the DC-link voltage (regulated to $V^*_{dclink\_2}$) is higher than its demand $V^*_{dclink\_1}$. It will then start to operate in buck mode, either in current mode or in voltage mode, depending on the battery voltage V_B. If it operates in current mode, the battery current demand is set to P^*_{Batt}/V_B where P^*_{Batt} is the battery charging power demand and $P^*_{Batt} = \text{Ramp}(|P_{ref}|)$, (the absolute value is used because the P_{ref} in this case is negative). The rate of change of the ramp function needs to be slow enough so it does not disturb the DC-link voltage controller. For instance, if the P^*_{Batt} is changed suddenly, the DC/DC converter will draw a large amount of power from the DC-link capacitor, which may result in a drastic drop in the DC-link voltage before the DC-link voltage controller reacts to this drop. To avoid any unnecessary transient, any changes required to P^* and Q^* values in the power flow controller also happen gradually via ramp functions.

9.3 Core Controller

Figure 9.3c shows the block diagram of one inverter phase and its core controller. The controller is implemented in the *abc* frame. The core voltage controller consists of an outer feedback loop of the capacitor voltage and an inner feedback loop of the capacitor current; the latter provides damping of filter resonance. A feedforward loop of the reference voltage is also implemented to improve the speed of response and minimize the steady-state error. Without the virtual impedance loop, the output voltage can be shown to be given by

$$V_o = G(s)V_c^* - Z(s)I_o \tag{9.1}$$

where $G(s)$ is the closed loop transfer function and $Z(s)$ is the system output impedance,

$$G(s) = \frac{k_v + 1}{L_1 C s^2 + k_c C s + k_v + 1} \tag{9.2}$$

$$Z(s) = \frac{L_1 s}{L_1 C s^2 + k_c C s + k_v + 1} + L_2 s \tag{9.3}$$

The voltage and current proportional gains k_v and k_c were chosen to be 2.0 and 2.2, respectively, to provide good transient and steady-state response (see Chapter 6). The step response settling time for $G(s)$ is about 1 ms and thus, from the point of view of the outer power flow loop, it will be assumed as an ideal voltage source with settable magnitude and frequency.

9.3.1 Virtual Impedance and Grid Harmonics Rejection

Virtual impedance has been proposed in the literature to guarantee a predominant inductive output impedance of the inverter, so that the active power is predominantly determined by the power angle and the reactive power is predominantly determined by the voltage amplitude [24, 25, 27–29]. In this UPS system, an LCL filter is used (rather than LC) with an extra grid side inductor L_2 to block the high switching frequency component of the output current from being injected into the grid. Obviously, replacing L_2 by a virtual impedance will not block the switching frequency current component. The LCL filter gives predominant inductive impedance around the fundamental frequency, which can arguably mean that the virtual impedance is not needed. The main motivation, however, for using the virtual impedance loop in this chapter is to improve the quality of the output current in grid-connected mode. During this mode, the inverter output current is determined by the two voltage sources: V and V_o and the impedance between them as shown in Figure 9.4. Thus, the harmonic content of the output current is determined by the grid voltage harmonics and the UPS output impedance. In stand-alone mode, however, high output impedance is not favored due to high voltage THD (total harmonic distortion) when supplying a nonlinear load. Therefore, the virtual impedance can be activated in grid-connected mode to improve current THD, and disabled in stand-alone mode so it does not affect the voltage THD. This is an important advantage of virtual impedance over physical impedance besides the obvious benefits of saving size and cost. The ideal transfer function of a virtual inductor L_v is given by

$$Z_{vi}(s) = sL_v \qquad (9.4)$$

However, implementing the derivative as in Equation 9.4 digitally introduces the well-known problem of noise amplification. Therefore, the pure derivative is replaced by a high-pass filter which is, in fact, a pure derivative preceded by a low pass filter. The transfer function of the virtual inductor is therefore given by

$$Z_v(s) = \frac{\omega_c s}{s + \omega_c} L_v \qquad (9.5)$$

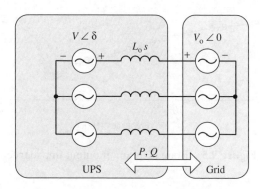

Figure 9.4 Equivalent circuit diagram of grid-connected UPS

where ω_c is the cut-off frequency of the high pass filter. When the virtual inductance is implemented, the output voltage can be shown to be given by

$$V_o = G(s)V_c^* - Z_o(s)I_o \tag{9.6}$$

where $Z_o(s)$ is the inverter modified output impedance and is given by

$$Z_o(s) = Z(s) + G(s)Z_v(s) \tag{9.7}$$

The Bode diagrams of both $Z(s)$ and $Z_o(s)$ for $L_v = 650\ \mu H$ are shown in Figure 9.5. By choosing $\omega_c = 1500\mathrm{rad\ s}^{-1}$, the virtual impedance provides higher impedance to the most significant grid harmonics (up to the 13th harmonic). As a result, this will increase the system's ability to reject grid voltage harmonics. The Bode diagram in Figure 9.5 shows predominant inductive impedance in the vicinity of the fundamental frequency. Thus, the output impedance (around the fundamental frequency) can be approximated as

$$Z_o(s) \approx L_o s \tag{9.8}$$

where L_o is the effective output inductance of the system that will be used in the analysis to follow. Its value from the Bode diagram of $Z_o(s)$ is $L_o = 996\ \mu H$.

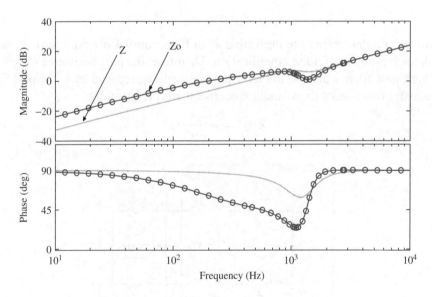

Figure 9.5   Bode diagram of output impedance

9.4 Power Flow Controller

The power flow controller that sets the voltage reference to the core controller has to fulfill the following tasks:

1. Equitable load-sharing between UPS (and other DG) units in stand-alone parallel mode.
2. Precise control of bidirectional power flow in grid-connected mode by each UPS unit.
3. Seamless transfer from grid-connected mode to stand-alone parallel mode and vice versa by the microgrid.

The proposed power flow controller was illustrated in Figure 9.3a. In grid-connected mode, when power is imported from the grid to charge the battery (if required), the active power demand is set by a voltage controller that regulates the DC-link voltage. In this case, the power demand is negative, which results in the power being imported from the grid. Hence, the converter operates as an active rectifier regulating the DC-link voltage. It is also possible to set the power to a positive value in order to export power into the grid. The DC-link voltage in this case is regulated by the DC/DC converter which now operates in boost mode, as discussed earlier.

9.4.1 Drooping Control Equations

The drooping control in stand-alone mode is given by

$$\omega = \omega_o^* - k_\omega(P - P^*) \tag{9.9}$$

$$V = V_0^* - k_a(Q - Q^*) \tag{9.10}$$

where ω_0^*, V_0^*, k_ω, and k_a are the nominal frequency reference, nominal voltage reference, proportional frequency drooping coefficient, and proportional voltage drooping coefficient, respectively. The values P^* and Q^* are set to nominal active and reactive power P_n and Q_n in stand-alone mode to improve frequency and voltage regulation. This is a bidirectional UPS with a maximum charging power of $-10\,\text{kW}$ (determined by the battery maximum charging current) and a maximum discharging power of $60\,\text{kW}$. The reactive power range is -45kVAR to $45\,\text{kVARA}$. Therefore, the nominal power settings P_n and Q_n are set to $25\,\text{kW}$ and zero, respectively, so the output frequency and voltage are within the allowed limit when the UPS produces maximum and minimum active and reactive power.

When the inverter is connected to a stiff grid, the frequency and voltage drooping technique needs to be modified to cope with grid voltage and frequency variations. Stiffness means that the voltage and frequency do not change by the behavior of small power sources. In the steady state, the inverter runs at the same frequency as that of the grid, $\omega = \omega_0$ (transient frequencies will be different). Similarly, the inverter voltage will nearly equal the grid voltage plus the small voltage drop across the inverter output impedance, that is, $V \approx V_0$. Assuming that the grid frequency and voltage equal their nominal values used by the controller such as $\omega_0 = \omega_0^*$ and $V_0 = V_0^*$, then according to Equations 9.9 and 9.10, $P = P^*$ and $Q = Q^*$ in the steady state. In this case, P^* and Q^* are set to active and reactive power commands P_{ref} and Q_{ref}, respectively. However, the grid frequency and voltage can vary with time due to variation of active and reactive loads, or the connection or disconnection of big generators. If the grid frequency and voltage do not equal their nominal reference values used by the inverter controller such as $\omega_0 \neq \omega_0^*$ and $V_0 \neq V_0^*$ then the UPS unit will produce active and reactive powers that are different from those of the set points. Typically, the grid frequency can change by 2% and the grid voltage can change by 10%, which results in significant deviations of the inverter's output active and reactive power from the set point values. To overcome this problem, the basic linear drooping controller is replaced by a proportional-integral (PI) controller to eliminate these power steady-state errors. The proposed drooping controller in grid-connected mode is given by

$$\omega = \omega_0^* - \left(k_\omega + \frac{k_{\omega\_I}}{s} \right)(P - P^*) \tag{9.11}$$

$$V = V_0^* - \left(k_a + \frac{k_{a\_I}}{s} \right)(Q - Q^*) \tag{9.12}$$

where k_{ω_I}, k_{a_I} are the integral frequency and voltage drooping coefficients, respectively. Usually the grid voltage and frequency do not vary very quickly and hence the time response of the integral parts of the PI controllers can be relatively slow, but it needs to be fast enough to track changes in grid frequency and voltage.

9.4.2 Small Signal Analysis

Figure 9.4 shows an equivalent circuit diagram of a grid-connected UPS unit. The grid impedance has been neglected as the UPS output impedance is usually much higher than the grid impedance. The active and reactive power flow between the UPS unit and the grid can be shown to be given by

$$P = \frac{3VV_0 \sin \delta}{\omega_0 L_0} \tag{9.13}$$

$$Q = \frac{3(VV_0 \cos \delta - V_0^2)}{\omega_0 L_0} \tag{9.14}$$

A small change in active power $\widetilde{P}$ is given by

$$\widetilde{P} = \frac{\partial P}{\partial \delta}\widetilde{\delta} + \frac{\partial P}{\partial V}\widetilde{V} \tag{9.15}$$

$$\widetilde{P} = 3\frac{V_0}{\omega_0 L_0}(\sin \delta_{eq} \cdot \widetilde{V} + V_{eq} \cos \delta_{eq} \cdot \widetilde{\delta}) \tag{9.16}$$

Similarly, a small change in reactive power $\tilde{Q}$ is given by

$$\tilde{Q} = \frac{\partial Q}{\partial \delta}\tilde{\delta} + \frac{\partial Q}{\partial V}\tilde{V} \tag{9.17}$$

$$\tilde{Q} = 3\frac{V_o}{\omega_o L_o}(\cos \delta_{eq} \cdot \tilde{V} - V_{eq} \sin \delta_{eq} \cdot \tilde{\delta}) \tag{9.18}$$

where δ_{eq} and V_{eq} are the equilibrium points around which the small signal analysis is performed. If the equilibrium points are chosen such that $\delta_{eq} = 0$ and $V_{eq} = V_o$, Equations 9.16 and 9.18 become

$$\tilde{P} = \frac{3V_o^2}{\omega_o L_o}\tilde{\delta} \tag{9.19}$$

$$\tilde{Q} = \frac{3V_o}{\omega_o L_o}\tilde{V} \tag{9.20}$$

By perturbing Equations 9.11 and 9.12 we get

$$\tilde{\omega} = \tilde{\omega}_o + \left(k_\omega + \frac{k_{\omega\_I}}{s}\right)(\tilde{P}^* - \tilde{P}_{meas}) \tag{9.21}$$

$$\tilde{V} = \tilde{V}_o + \left(k_a + \frac{k_{a\_I}}{s}\right)(\tilde{Q}^* - \tilde{Q}_{meas}) \tag{9.22}$$

where $\tilde{P}_{meas}$ and $\tilde{Q}_{meas}$ are small changes in the measured active and reactive power, respectively. The grid frequency and voltage reference points ω_o^* and V_o^* are fixed by the controller. However, the grid frequency ω_o and voltage V_o may change slightly and hence the notations $\tilde{\omega}_o$ and $\tilde{V}_o$ in Equations 9.21 and 9.22 represent the small deviations in the grid frequency and voltage from their nominal values, respectively. The small signal model can be represented by a block diagram, as shown in Figure 9.6 where the angle δ_o is the initial angle between the inverter voltage and the common AC bus voltage before connection. It is normally caused by measurement error and it is represented as a disturbance in the block diagram.

The transfer function $F(s)$ in Figure 9.6 is the power measurement transfer function that relates the measured power to the actual instantaneous power. In a balanced three-phase system, the instantaneous total power is free from ripple. However, if the system is unbalanced, undesired ripple will appear in the instantaneous total power signal and a filter $F(s)$ has to be used. $F(s)$ can be a simple low pass filter that is,

$$F(s) = \frac{\omega_f}{s + \omega_f} \tag{9.23}$$

The cut-off frequency of the low pass filter ω_f needs to be low enough to filter out the undesired terms but it needs to be high enough to give an adequately fast power

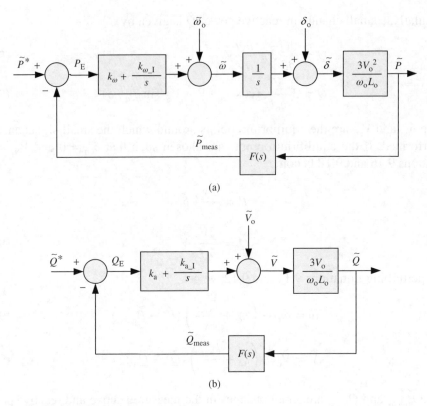

Figure 9.6   Small signal model, (a) active power and (b) reactive power

measurement response time. It is common practice to select the cut-off frequency to be one decade below the fundamental frequency [25]. Alternatively, the average of the power can be calculated by integrating the instantaneous power over one fundamental cycle T using the following equations,

$$P_{\text{meas}} = \frac{1}{T} \int_{t-T}^{t} p(t)\mathrm{d}t \qquad (9.24)$$

$$Q_{\text{meas}} = \frac{1}{T} \int_{t-T}^{t} q(t)\mathrm{d}t \qquad (9.25)$$

The integration is rolled over the last fundamental cycle and is updated every time new voltage and current samples are available. In this case, the power measurement transfer function, hereafter called the sliding filter, can be given by

$$F(s) = \frac{1}{Ts}(1 - \mathrm{e}^{-Ts}) \qquad (9.26)$$

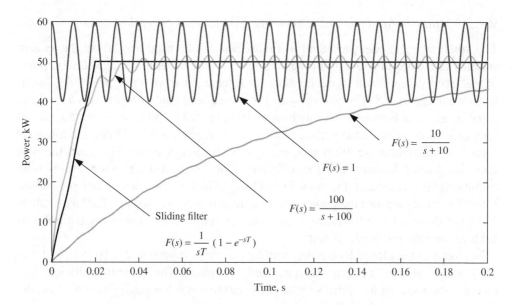

Figure 9.7   Response of power calculation methods

The power measured using the sliding filter does not suffer from signal ripple and it has a fast response time. Figure 9.7 shows the response of the two power measurement filters of a three-phase unbalanced system where one of the phase loads is 50% of the loads on the other two phases. Without using any kind of filtration, that is, $F(s) = 1$, the instantaneous power has a large ripple component. When a low pass filter with a cut-off frequency of $100 \, \text{rad s}^{-1}$ is used, the time response is fast but the power measurement suffers from a steady state ripple. When the cutoff frequency is reduced to $10 \, \text{rad s}^{-1}$, the ripple is attenuated but the time response is quite sluggish. The advantage of using the sliding filter over the low pass filter is clear; it gives a ripple free signal with faster response.

The time delay in Equation 9.26 can by approximated by a rational function using the Padé approximation. If a second order Padé approximation is used, the time delay can be expressed as

$$e^{-sT} \approx \frac{T^2/12\,s^2 - T/2\,s + 1}{T^2/12\,s^2 + T/2\,s + 1} \tag{9.27}$$

Substituting Equation 9.27 in Equation 9.26 gives

$$F(s) \approx \frac{1}{T^2/12\,s^2 + T/2\,s + 1} \tag{9.28}$$

It is worth noting that Equation 9.28 represents a second order low pass filter with a cut-off frequency of $140 \, \text{rad s}^{-1}$.

9.4.3 Stability Analysis and Drooping Coefficients Selection

The small signal model presented in Figure 9.6 can be used to analyze the system stability and determine the drooping coefficients. The locus of the closed loop poles of the active power small signal model with k_ω changes from 0 to $7 \times 10^{-4} (k_{\omega\_I} = 0)$ is shown in Figure 9.8. The poles $\gamma_{1,2}$ are the closed loop poles when $F(s)$ is a low pass filter, as given in Equation 9.23 with $\omega_f = 10$ rad s^{-1}. The poles $\lambda_{1,2,3}$ are the closed loop poles when $F(s)$ is the sliding filter given in Equation 9.28. When the low pass filter is used to measure the power, the system is always stable. However, the root locus has a fixed distance from the imaginary axis which means that the settling time of the system is fixed and it is not affected by k_ω. On the other hand, when the sliding filter is used to measure the power, the system can become unstable for large values of k_ω but there is more freedom in choosing the real value of the dominant poles to achieve the required speed of response.

The locus of the closed loop poles with k_{ω_I} changes from 0 to $2 \times 10^{-4} (k_\omega = 1.5 \times 10^{-4})$ is also shown in Figure 9.8 (enclosed by circles). The integral coefficient k_{ω_I} changes the locus of the poles very slightly and hence it has negligible effect on the dynamic response of the system.

Figure 9.9 shows the root locus of the reactive power small signal model with k_a changing from 0 to 6×10^{-4} ($k_{a\_I} = 0$). For both power measurement filters, the system is always stable. However, when the sliding filter is used, the system step response

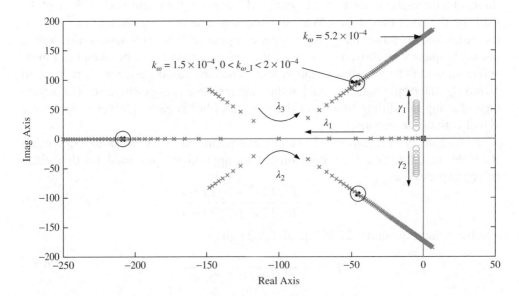

Figure 9.8 Root locus diagram of the active power model, $0 < k_\omega < 7 \times 10^{-4}$, $k_{\omega\_I} = 0$, $\gamma_{1,2}$ are the poles when LPF ($\omega_f = 10$ rad s^{-1}) is used, $\lambda_{1,2,3}$ are the poles when a sliding filter is used. The poles enclosed by the circles represent the root locus with varying integral coefficient $0 < k_{\omega\_I} < 2 \times 10^{-4}$, $k_\omega = 1.5 \times 10^{-4}$

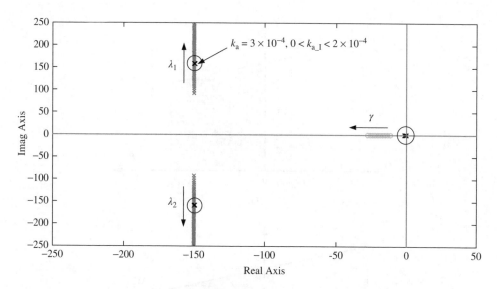

Figure 9.9 Root locus diagram of the reactive power model, $0 < k_a < 6 \times 10^{-4}$, $k_{a\_I} = 0$, γ is the pole when LPF ($\omega_f = 10 \text{ rad s}^{-1}$) is used, $\lambda_{1,2}$ are the poles a when sliding filter is used. The poles enclosed by the circles represent the root locus with varying integral coefficient $0 < k_{a\_I} < 2 \times 10^{-4}$, $k_a = 3 \times 10^{-4}$

is under-damped compared to the exponential response obtained when the LPF is used. In addition, the closed loop poles are located further to the left, which results in a faster response. The locus of the poles for varying k_{a_I} (enclosed by circles) shows that it has negligible effect on the dynamic response of the system.

The frequency drooping coefficient k_ω needs to be carefully chosen to satisfy transient and steady-state requirements. In terms of the steady state, the maximum frequency deviations caused by the drooping control in stand-alone mode should not exceed the maximum allowed limits. According to Equation 9.9, the maximum frequency drooping coefficient is given by

$$k_{\omega\_max} = \frac{2\pi \Delta f_{max}}{P_{max} - P_n} \tag{9.29}$$

The above equation is illustrated in Figure 9.10a. For $\Delta f_{max} = 1 \text{ Hz}$, $P_{max} = 60 \text{ kW}$, $P_n = 25 \text{ kW}$, the maximum frequency drooping coefficient is $k_{\omega\_max} = 1.8 \times 10^{-4}$. According to the root locus in Figure 9.8, the system is stable for this gain value.

In terms of transient response, the system should have a good damping ratio, ideally between 0.3 and 0.7. A number of different variables can also be examined to analyze the transient response of the controller, such as circulating current as in [25]. In this design, the amount of energy that can be transferred from the AC bus to the UPS unit during a grid connection transient is examined. If the angle of the UPS output voltage lags the AC bus voltage by δ_o when it first connects to the AC bus, active power will

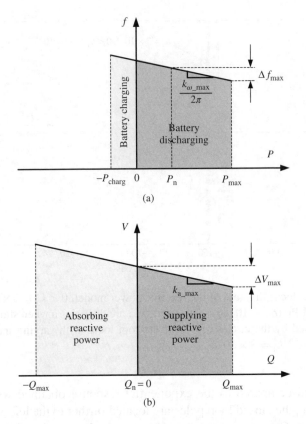

Figure 9.10 Droop characteristic, (a) frequency and (b) voltage

flow from the AC bus to the UPS and the energy transferred will cause the DC-link voltage to rise. If the DC-link voltage is higher than the maximum limit, the protection system will trip. The trip limit V_{DC_max} is specified according to the voltage rating of the power switches of the converter. The objective is to select the drooping gain so that the maximum transient energy absorbed by the converter (caused by the presence of an initial power angle δ_o) does not cause the DC-link voltage to rise above the trip limit. If the demand power $\widetilde{P}^*$ is set to zero in Figure 9.6a, and by ignoring the disturbance caused by the frequency deviation, the block diagram that relates the output energy (integral of power) to the disturbance δ_o is shown in Figure 9.11. The proportional coefficient k_ω needs to be chosen to limit the overshoot in the output energy but without compromising relative stability. The integral coefficient k_{ω_I} does not affect the transient response, as was shown earlier in Figure 9.8. Table 9.2 summarizes the results of the damping ratio and the overshoot in the output energy for different values of k_ω. Increasing k_ω reduces the overshoot in the output energy but also reduces the damping ratio which means that a compromise has to be made. The superiority of the sliding filter over the low pass filter can also be seen from the results in Table 9.2.

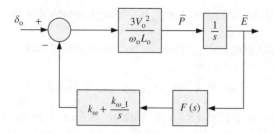

Figure 9.11   Block diagram relating $\widehat{E}$ to δ_o

Table 9.2   Transient response

k_ω	Low pass filter		Sliding filter	
	ζ	Energy overshoot (J)	ζ	Energy overshoot (J)
1.0×10^{-4}	0.22	$25000 \times \delta_o$	0.65	$11000 \times \delta_o$
1.5×10^{-4}	0.18	$20000 \times \delta_o$	0.44	$8200 \times \delta_o$
2.0×10^{-4}	0.16	$17000 \times \delta_o$	0.32	$7000 \times \delta_o$

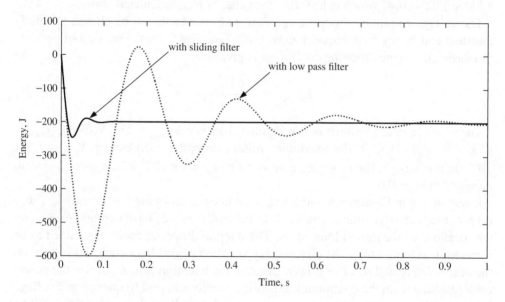

Figure 9.12   Energy transient response, $\delta_o = -0.03$ rad, $k_\omega = 1.5 \times 10^{-4}$

For example, for $k_\omega = 1.5 \times 10^{-4}$, the damping ratio, when the sliding filter is used, is about two and a half times more and the overshoot is 60% less than that when the low pass filter is used. Figure 9.12 shows the transient response of the output energy for an initial power angle of $\delta_o = -0.03$ rad for the two different methods of power measurement. The negative sign of the power angle means that the UPS output voltage lags the AC bus and the negative sign of the energy means that the energy is absorbed by the UPS unit.

The maximum transient energy that the converter can absorb is given by

$$E_{max} = \frac{1}{2}C_{DC}\left(V^2_{DClink\_max} - V^2_{DClink\_o}\right) \tag{9.30}$$

where V_{DClink_o} is the initial DC-link voltage before transient, V_{DClink_max} is the maximum allowed DC-link voltage, and C_{DC} is the DC-link capacitance. According to the values in Table 9.1, and knowing that when the inverter first connects to the AC bus, the initial DC-link voltage is $V_{DClink\_o} = 750V$, the maximum energy according to Equation 9.30 is $E_{max} = 438J$. In order to prevent the DC-link voltage from rising to the maximum limit of 1000 V, the overshoot transient energy should be less than E_{max}. The initial angle δ_0 depends on the measurement error in the synchronization method (phase-locked loop or zero crossing detection). The synchronization in this design is performed by detecting the zero crossing of the AC bus voltage. The error in this case is limited to one sampling period T_s of the AC bus voltage, which for the sampling frequency of 16 kHz is equal to 0.02 rad. The drooping coefficient is set to $k_\omega = 1.5 \times 10^{-4}$. According to Table 9.2, the maximum energy overshoot is $8200 \times 0.02 = 164J$ which is less than the value of E_{max} calculated above.

The voltage amplitude drooping coefficients k_a needs also to be chosen to satisfy transient and steady state requirements. From Equation 9.10, the maximum possible proportional voltage drooping coefficient is given by

$$k_{a\_max} = \frac{\Delta V_{max}}{Q_{max} - Q_n} \tag{9.31}$$

The above equation is illustrated in Figure 9.10b. For $\Delta V_{max} = 23.0$ V (rms), $Q_{max} = 45$ kVAR, $Q_n = 0$ kVAR, the maximum voltage drooping coefficient is $k_{a_{max}} = 5.1 \times 10^{-4}$. In this design, the drooping gain is set to $k_a = 3 \times 10^{-4}$ which gives a system damping ratio of 0.6.

It was shown in Figures 9.8 and 9.9 that the integral drooping coefficients k_{ω_I}, k_{a_I} do not change the dynamic response of the controller as they have negligible effect on the locations of the closed loop poles. The integral drooping coefficients need to be selected so as to minimize the effect of any variation in grid frequency and voltage on the output active and reactive power. Because the intention is to minimize the power error resulting from the continuous dynamic deviations in grid frequency and voltage (not only against static deviations), in the analysis to follow, these deviations will be modeled as ramp functions as follows:

$$\widetilde{\omega}_o = \frac{D_\omega}{s^2} \tag{9.32}$$

$$\widetilde{V}_o = \frac{D_a}{s^2} \tag{9.33}$$

where D_ω is the rate of change of the frequency deviation in rad s^{-2} and D_a is the rate of change of the voltage deviation in V s^{-1}. From Figure 9.6a, the error signal of the active power P_E can be shown to be given by

$$P_E = \frac{(-3V_o^2/\omega_o L_o)s}{\frac{T^2}{12}s^4 + \frac{T}{2}s^3 + s^2 + k_\omega \frac{3V_o^2}{\omega_o L_o}s + k_{\omega\_I}\frac{3V_o^2}{\omega_o L_o}}\tilde{\omega}_o \tag{9.34}$$

The bandwidth of the above transfer function is about 16 Hz, which is much higher than the frequency of change of voltage and grid frequency. The steady-state error near the end of the ramp can be estimated using the final-value theorem. Substituting Equation 9.32 in Equation 9.34, the steady-state error in active power P_{E_ss} is given by

$$P_{E\_ss} = \lim_{s\to 0}(sP_E) = \frac{-D_\omega}{k_{\omega\_I}} \tag{9.35}$$

Similarly, from Figure 9.6b the error signal of the reactive power Q_E can be shown to be given by

$$Q_E = \frac{(-3V_o/\omega_o L_o)s}{\frac{T^2}{12}s^3 + \frac{T}{2}s^2 + \left(k_a\frac{3V_o^2}{\omega_o L_o}+1\right)s + k_{a\_I}\frac{3V_o}{\omega_o L_o}}\tilde{V}_o \tag{9.36}$$

substituting Equation 9.33 into Equation 9.36, the steady-state error in reactive power Q_{E_ss} is given by

$$Q_{E\_ss} = \lim_{s\to 0}(sQ_E) = \frac{-D_a}{k_{a\_I}} \tag{9.37}$$

The coefficients k_{ω_I} and k_{a_I} are set to 5×10^{-5} and 1×10^{-4}, respectively. It has been found that at the test site, the grid can drift by up to 0.05 Hz and 4 V in 5 min. This corresponds to $D_\omega = 0.001 \mathrm{rads}^{-2}$ and $D_a = 0.013 \mathrm{Vs}^{-1}$. Therefore, according to Equations 9.35 and 9.37, the steady-state errors in active and reactive power caused by continuous dynamic deviation of the grid frequency and voltage are only $P_{E\_ss} = \pm 20\mathrm{W}$ and $Q_{E\_ss} = \pm 130\mathrm{VAR}$, respectively. It is important to note that these errors only appear during the dynamic deviation of voltage and/or frequency. Once the deviation has stopped, the integral gains bring the steady-state errors to zero.

To validate the small signal model, a detailed model of a three-phase half-bridge PWM inverter with LCL filter, as per Tables 9.1 and 9.3, was built in MATLAB®/Simulink. The power was measured in the model using the sliding filter. When the inverter connects to the grid, the grid voltage signals lead the inverter voltage signals (measured at the filter capacitors) by 0.039 rad. Figure 9.13 shows the response of the inverter transient absorbed energy compared to that of the small signal model produced using Figure 9.6a. It can be seen that the small signal model provides a reasonable prediction of the response of the system in terms of overshoot and frequency of oscillations, given the assumptions made to derive the small signal model, such as assuming the inverter as an ideal voltage source.

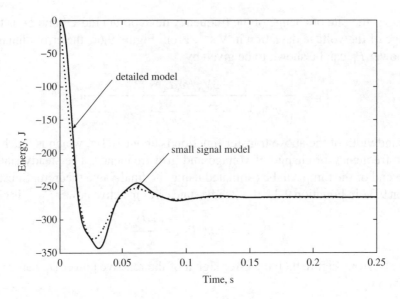

Figure 9.13   Energy transient response, actual model and small signal model $\delta_o = 0.039\text{rad}$

9.5   DC Link Voltage Controller

The DC-link voltage controller regulates the DC-link voltage during the battery charging mode. A PI controller is chosen and the overall block diagram of the DC-link voltage controller, including the small signal model of the inner loop power flow controller (ignoring the frequency and angle disturbances), is shown in Figure 9.14. The power drawn by the DC/DC converter is represented by $\widetilde{P}_{\text{Batt}}$. It can be noticed that the model is nonlinear due to the presence of the square root function. In order to be able to use linear control design techniques, the square root relation is linearized. Let $x = V_{\text{DClink}}^2$, and $y(x) = V_{\text{DClink}} = \sqrt{x}$, a small change in y is given by

$$\widetilde{y} = \widetilde{x}\left.\frac{dy}{dx}\right|_{x=x_o} \tag{9.38}$$

where $\widetilde{x}$ is a small change in x, and x_o is the point around which the linearization is performed. Given that the DC-link voltage range of interest is from 750 to 800 V, the linearization point is chosen to be $x_o = 775^2$,

$$\widetilde{y} = \widetilde{x}\left.\frac{1}{2\sqrt{x}}\right|_{x=775^2} \tag{9.39}$$

Hence, a small change in the DC-link voltage can be written as

$$\widetilde{y} = m\widetilde{x} \tag{9.40}$$

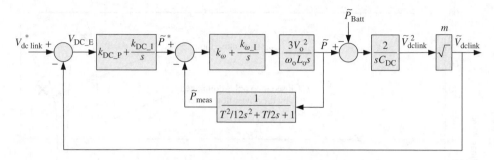

Figure 9.14   DC link voltage controller

where $m = \dfrac{1}{2\sqrt{x}}\bigg|_{x=775^2} = 6.5 \times 10^{-4}$.

The open loop transfer function that relates $\widetilde{V}_{\text{DClink}}$ to $\widetilde{P}^*$ (assuming $\widetilde{P}_{\text{Batt}} = 0$) is given by

$$G_{\text{DC}}(s) = \frac{2mA}{C_{\text{DC}}} \left(\frac{\left(\frac{T^2}{12}s^2 + \frac{T}{2}s + 1 \right)(k_\omega s + k_{\omega\_\text{I}})}{\frac{T^2}{12}s^5 + \frac{T}{2}s^4 + s^3 + Ak_\omega s^2 + Ak_{\omega\_\text{I}}s} \right) \tag{9.41}$$

where $A = 3V_o^2/\omega_o L_o$.

The design of the DC-link voltage controller now becomes straightforward. The PI gains are chosen so that the dynamic response of the DC-link controller is much slower than the inner power flow controller in order to decouple the two controllers. The proportional and integral gains $k_{\text{DC\_P}}$ and $k_{\text{DC\_I}}$ are chosen to be 40 and 2000, respectively. Figure 9.15 shows the closed loop poles of the complete system. The high frequency poles are the ones caused by the power flow loop, whereas the dominant lower frequency poles are caused by DC-link voltage loop; the dynamics of the dominant closed loop poles are well decoupled from the dynamics of the power flow closed loop poles. The simulated response of the controller to step the voltage up from the boost regulated value of 750 V to the active rectifier demand value of 800 V is shown in Figure 9.16. The rise and settling times are 0.05 and 0.3 s, respectively.

The block diagram in Figure 9.14 can be used to determine the ramp rate of the power demand sent to the DC/DC (during battery charging mode) to produce P^*_{Batt}, (see Figure 9.3b). The ramp rate should be slow enough so that it does not cause any major disturbance on the DC-link voltage. For example, if a sudden power is drawn from the DC-link capacitor by the DC/DC converter, the capacitor voltage might drop significantly before the DC-link voltage controller reacts to the disturbance. The power drawn by the DC/DC converter is modeled as a ramp function such as

$$\widetilde{P}_{\text{Batt}} = \frac{D_{\text{P}}}{s^2} \tag{9.42}$$

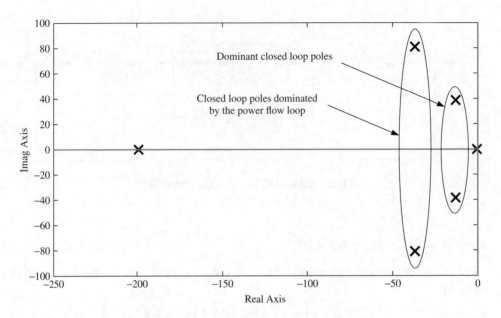

Figure 9.15   Closed loop poles of the system

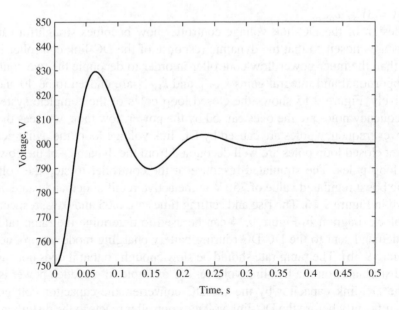

Figure 9.16   Step response of DC-link voltage controller

where D_P is the rate of change in W s^{-1}. From Figure 9.14 and by using the final-value theorem, the steady-state error in V_{DClink} as a result of the dynamic change in P^*_{Batt} can be shown to be given by

$$V_{DC\_Ess} = \lim_{s \to 0} (s V_{DC\_E}) = \frac{-D_P}{k_{DC\_I}} \qquad (9.43)$$

The ramp rate D_P is set to $1\,\mathrm{kW\,s^{-1}}$. Therefore, the steady-state error caused by ramping up/down the DC/DC power demand is only $V_{DC\_Ess} = \pm 2\mathrm{V}$.

9.6 Experimental Results

A complete microgrid system was built and tested experimentally. The experimental set-up is illustrated in Figure 9.17. It includes two 60 kW line interactive UPS systems each supplied by a 30 A h (550–700 V) lithium-ion battery, and a STS with a supervisory controller. In addition, a 60 kW resistive load is used as a local load. The controller parameters of the converter are shown in Table 9.3. The controllers were implemented using the Texas Instrument TMS320F2812 Fixed point DSP (digital signal processor). In the practical implementation of the core controller, the value of the DC-link voltage needs to be taken into account when the internal counter that represents the PWM carrier is scaled. Because the DC-link voltage can have two values (750 V in battery discharging mode or 800 V in battery charging mode), the scaling factor needs to be updated accordingly. The low communication link between the STS and the UPS units was realized using the controller area network (CAN) protocol. The power demand references were set by the supervisory controller of the STS and sent via CAN bus. Also, for the purpose of testing different scenarios, the power demand could be set by the user via a CAN bus-connected computer. An update signal of the status of the STS is also sent to the UPS units via CAN bus. A picture of one of the UPS units is shown in Figure 9.18.

Figure 9.19 shows the grid-connected to stand-alone mode transition of one UPS unit. The UPS unit was operating in grid-connected mode, charging the battery at a 1 kW rate (note that the current is 180° out of phase with respect to the voltage, which means that the UPS is importing power from the grid). A 60 kW local resistive load was also connected to the AC bus and supported by the grid. When the grid fails, the UPS moves seamlessly to stand-alone mode and starts supplying the local load.

It can also be observed from the figure that at the moment when the grid fails, the DC-link voltage starts to drop from the active rectifier reference ($V^*_{DClink\_2} = 800\mathrm{V}$) due to the change of power flow direction. However, when the DC-link drops below the boost reference ($V^*_{DClink\_1} = 750\mathrm{V}$) the boost starts to regulate the DC-link voltage and raises it to 750 V.

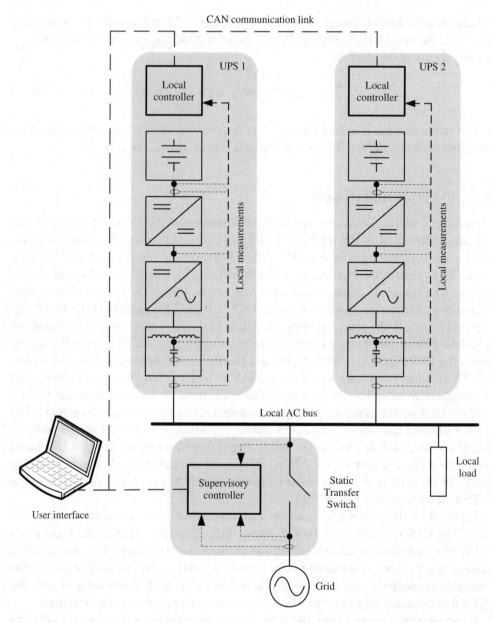

Figure 9.17 Experimental set-up

Figure 9.20 shows the grid-connected to stand-alone mode transition of the two UPS units. Each unit was producing 0 kW as both batteries were fully charged. At the moment when the grid fault occurs, the two units move seamlessly to stand-alone parallel mode sharing the 60 kW local load equally. The critical load does not see

Table 9.3 Controller parameter values

Symbol	Value	Description
f_{sw}	8 kHz	Inverter switching frequency
f_s	16 kHz	Voltage controller sampling frequency
k_v	2.0	Capacitor voltage loop proportional gain
k_c	2.2	Capacitor current loop proportional gain
k_ω	1.5×10^{-4}	Drooping frequency coefficient
k_a	3.0×10^{-4}	Drooping voltage coefficient
k_{ω_I}	5.0×10^{-5}	Integral drooping frequency coefficient
k_{a_I}	1.0×10^{-4}	Integral drooping voltage coefficient
ω_o^*	314.16 rad s^{-1}	Nominal frequency
V_o^*	230 V (rms)	Nominal grid voltage
k_{DC_P}	40	DC-link voltage controller proportional gain
k_{DC_I}	2000	DC-link voltage controller (active rectifier) integral gain
$V_{DClink\_1}^*$	750 V	Boost voltage demand
$V_{DClink\_2}^*$	800 V	Active rectification demand
P_{ref}	−10 to 60 kW	Active power demand from the supervisory controller
Q_{ref}	−45 to 45 kVAR	Reactive power demand from the supervisory controller
P_n	25 kW	Nominal active power rating demand in stand-alone mode
Q_n	0 kVAR	Nominal reactive power rating demand in stand-alone mode
L_v	650 μH	Virtual sharing inductor
ω_c	1500 rad s^{-1}	Cut-off frequency of the virtual impedance high pass filter

any power interruption, as can be seen from the load current signal. The two UPS units behave exactly the same, as can be seen from their current signals which are placed on top of each other. In Figure 9.21, the two UPS units were operating in grid-connected mode, charging their batteries at a 10 kW rate. When the grid fails, the two inverters move almost seamlessly from grid-connected mode to stand-alone paralleling mode and start sharing the critical load equally. There is, however, a small change in the current waveform due to the small change in the AC bus voltage. This is because the currents of the two UPS systems changes direction suddenly which creates a small voltage disturbance proportional to $L_o di_o/dt$. The UPS charging current looks quite distorted and this is because each unit generates only 17% of its rated power. According to the IEEE Standard 1547 [42], the THD limit is defined based on the unit maximum current, and the distortion at low power is therefore acceptable.

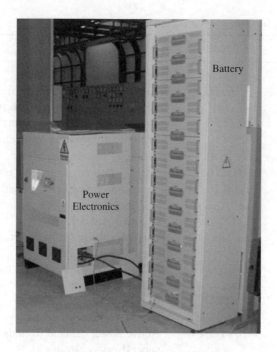

Figure 9.18   Line interactive UPS

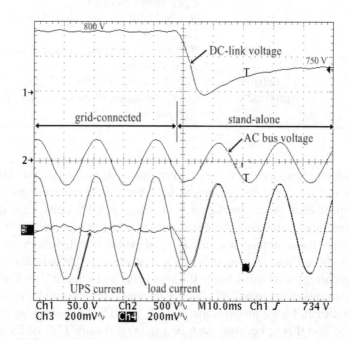

Figure 9.19   Grid-connected to stand-alone transition of one UPS unit. When in grid-connected mode, the UPS was charging the battery at 1 kW rate. When in stand-alone mode, the UPS unit was supplying a local 60 kW load. (Current scale: 1 A/2 mV)

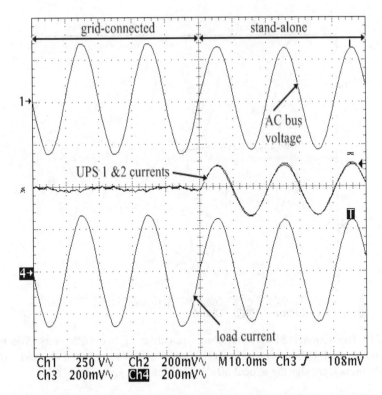

Figure 9.20 Grid-connected to stand-alone transition of two UPS units. When in grid-connected mode, each UPS was producing 0 kW. When in stand-alone mode, the two UPS units were sharing a local 60 kW load (current scale: 1 A/2 mV)

Figure 9.22 shows the transition response from stand-alone to grid-connected mode. During stand-alone mode, the microgrid was formed by one UPS unit supplying a 60 kW resistive load. When the grid becomes available, the STS closes and connects the microgrid seamlessly to the grid. Only small fluctuations appear on the DC-link which settles within two fundamental cycles. The load current sees a small increase because the grid voltage is slightly higher than the stand-alone UPS voltage because of the inherent characteristic of the voltage droop control. The UPS current starts to decrease and will eventually become equivalent to the power command in grid-connected mode.

Figure 9.23 shows the transient response when the active power demand P_{ref} changes from -10 to 60 kW by the user interface. The controller ramps up the power demand gradually (to avoid any unnecessary transient, as discussed earlier) and thus the current changes amplitude and phase gradually. The DC-link voltage drops from 800 V (controlled by the AC/DC active rectifier) to 750 V (controlled by the DC/DC converter). Figure 9.24 shows the DC-link voltage controller transient response during the transition from battery discharging mode (the DC-link voltage is controlled by

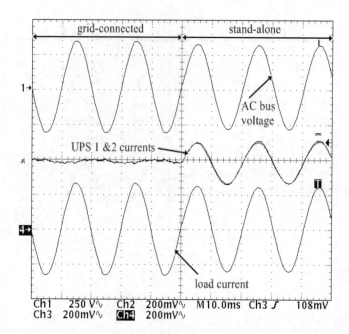

Figure 9.21 Grid-connected to stand-alone transition of two UPS units. When in grid-connected mode, each UPS was charging its battery at 10 kW rate. When in stand-alone mode, the two UPS units were sharing a local 60 kW load (current scale: 1 A/2 mV)

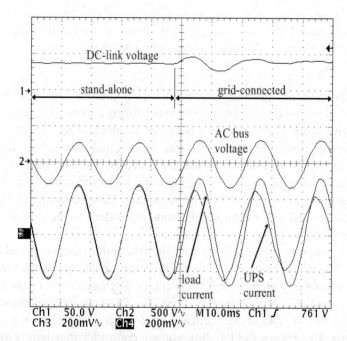

Figure 9.22 Stand-alone to grid-connected transition of one UPS unit, when in stand-alone mode, the UPS unit was supplying a local 60 kW load (current scale: 1 A/2 mV)

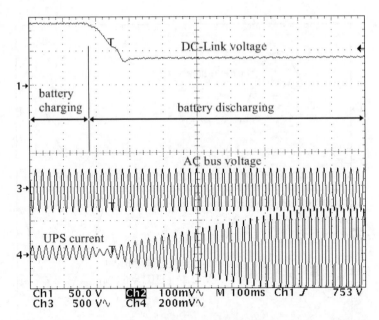

Figure 9.23   Battery charging (10 kW) to discharging (60 kW) mode transition (current scale: 1 A/2 mV)

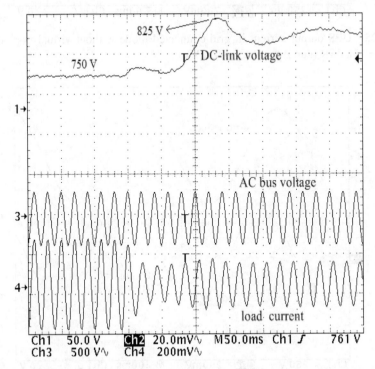

Figure 9.24   DC-link voltage controller transient response when going from battery discharging mode (the DC-link voltage is controlled by the DC/DC converter) to battery charging mode (the DC-link voltage is controlled by the AC/DC converter) (current scale: 1 A/2 mV)

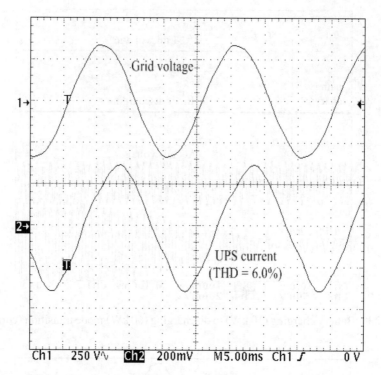

Figure 9.25 UPS output current in grid-connected mode without virtual impedance, $P =$ 60 kW, $Q = 45$ kVAR (current scale: 1 A/2 mV)

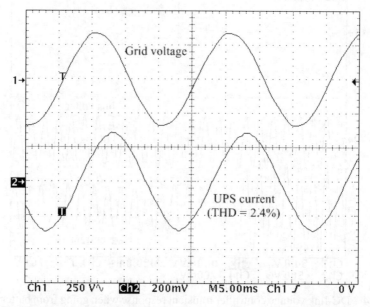

Figure 9.26 UPS output current in grid-connected mode with virtual impedance, $P = 60$ kW, $Q = 45$ kVAR (current scale: 1 A/2 mV)

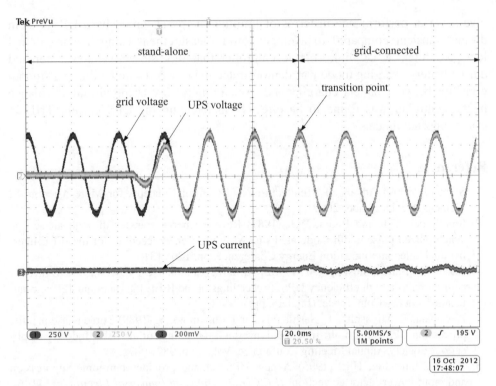

Figure 9.27 UPS and grid voltages during transition from stand-alone mode to grid-connected mode

the DC/DC converter) to battery charging mode (the DC-link voltage is controlled by the AC/DC converter). The DC-link shows a good transient response, similar to the simulated response shown in Figure 9.16. Figure 9.25 shows the bus voltage and output current in grid-connected mode. The UPS is supplying active power of 60 kW and reactive power of 45 kVAR. The virtual impedance is disabled and the current THD is measured to be 6.0%. In Figure 9.26 the virtual impedance is enabled and the THD dropped to only 2.4%. Figure 9.27 shows the starting sequence of one UPS unit in grid-connected mode. Initially, the inverter voltage is controlled to have the same magnitude and frequency as the grid voltage and it is also synchronized so the power angle is minimized to be virtually zero. When the STS controller is satisfied that the two voltage signals across its terminals are healthy and in phase it closes the STS so the microgrid is connected to the main grid.

9.7 Conclusions

A control strategy for a line interactive UPS system that is based on a voltage and frequency drooping technique was demonstrated experimentally to be capable of achieving seamless transfer between grid-connected and stand-alone parallel modes of operation. Measuring the average power using real time integration was found to give superior performance in terms of limiting energy absorbed by a UPS system during transient compared to average power measurement using a conventional low pass filter. A DC-link voltage controller that sets the active power demand during battery charging mode was demonstrated to be effective in facilitating smooth transition between battery charging and discharging modes. Using virtual impedance in the controller was found to be effective in reducing the UPS current THD in grid-connected mode.

References

1. Siemens (2009) Advanced Architectures and Control Concepts for More Microgrids, Siemens AG, December 2009.
2. Smollinger, G.J. and Raddi, W.J. (1981) Reverse energy transfer through an ac line synchronized pulse width modulated sine wave inverter. Proceedings of the IEEE International Telecommunication Energy Conference, pp. 126–131.
3. Raddi, W.J. and Johnson, R.W. (1982) An utility interactive PWM sine wave inverter configured as a high efficiency UPS. Proceedings of the IEEE International Telecommunication Energy Conference, (Intelec), pp. 42–48.
4. Kawabata, T., Miyashita, T., Sashida, N., and Yamamoto, Y. (1989) Three phase parallel processing UPS using multi-functional inverter. Proceedings of the IEEE Industry Applications Society Annual Meeting Conference, Vol. 1, pp. 982–987.
5. Wu, J.C. and Jou, H.L. (1995) A new UPS scheme provides harmonic suppression and input power factor correction. *IEEE Transactions on Industrial Electronics*, **42** (6), 629–635.
6. Joos, G., Lin, Y., Ziogas, P.D., and Lindsay, J.F. (1992) An online UPS with improved input-output characteristics. Proceedings of the IEEE Applied Power Electronics Conference and Expo, February 1992, pp. 598–605.
7. Lin, Y., Joos, G., and Lindsay, J.F. (1993) Performance analysis of parallel processing UPS systems. Proceedings of the IEEE Applied Power Electronics Conference and Expo, March 1993, pp. 533–539.
8. Rathmann, S. and Warner, H.A. New generation UPS technology, the delta conversion principle. Conference Proceedings of the 31st IEEE Industry Applications Society Annual Meeting, October 1996, Vol. 4, pp. 2389–2395.
9. Kamran, F. and Habetler, T.G. (1998) A novel on-line UPS with universal filtering capabilities. *IEEE Transactions on Power Electronics*, **13** (3), 410–418.
10. Jeon, S.J. and Cho, G.H. (1997) A series-parallel compensated uninterruptible power supply with sinusoidal input current and sinusoidal output voltage. Proceedings of the 28th IEEE Power Electronics Specialists Conference, June 1997, Vol. 1, pp. 297–303.
11. da Silva, S.A.O., Donoso-Garcia, P.F., Cortizo, P.C., and Seixas, P.F. (2002) A three-phase line-interactive UPS system implementation with series-parallel active power-line conditioning capabilities. *IEEE Transactions on Industry Applications*, **38** (6), 1581–1590.
12. Kwon, B.H., Choi, J.H. and Kim, T.W. (2001) Improved single-phase line-interactive UPS. *IEEE Transactions on Industrial Electronics*, **48** (4), 804–811.
13. Jou, H.-L., Wu, J.-C., Tsai, C. *et al.* (2004) Novel line-interactive uninterruptible power supply. Proceedings of the Institution of Electrical Engineers, Electric Power Applications, Vol. 151, no. 3, pp. 359–364.

14. Barrero, F., Martinez, S., Yeves, F. *et al.* (2003) Universal and reconfigurable to UPS active power filter for line conditioning. *IEEE Transactions on Power Delivery*, **18** (1), 283–290.

15. Yeh, C.-C. and Manjrekar, M.D. (2007) A reconfigurable uninterruptible power supply system for multiple power quality applications. *IEEE Transactions on Power Electronics*, **22** (4), 1361–1372.

16. Tirumala, R., Mohan, N., and Henze, C. (2002) Seamless transfer of grid connected PWM inverters between utility-interactive and stand-alone modes. Proceedings of the IEEE APEC, Dallas, TX, March 2002, pp. 1081–1086.

17. Pai, F.S. and Huang, S.J. (2006) A novel design of line-interactive uninterruptible power supplies without load current sensors. *IEEE Transactions on Power Electronics*, **21** (1), 202–210.

18. Tao, H., Duarte, J.L. and Hendrix, M.A.M. (2008) Line-interactive UPS using fuel cell as the primary source. *IEEE Transactions on Industrial Electronics*, **55** (8), 3005–3011.

19. Arias, M., Fernandez, A., Lamar, D.G. *et al.* (2008) Simplified voltage-sag filler for line-interactive uninterruptible power supplies. *IEEE Transactions on Industrial Electronics*, **55** (8), 3005–3011.

20. Kim, H., Yu, T. and Choi, S. (2008) Indirect current control algorithm for utility interactive inverters in distributed generation systems. *IEEE Transactions on Power Electronics*, **23** (3), 1342–1347.

21. Ho, W.J., Lio, J.B., and Feng, W.S. (1997) A line-interactive UPS structure with built-in vector-controlled charger and PFC. Proceedings of the International Conference on Power Electronics and Drive Systems, pp. 127–132.

22. Okui, Y., Ohta, S., Nakamura, N. *et al.* (2003) Development of line interactive type UPS using a novel control system. Proceedings of the IEEE International Telecommunication Energy Conference, pp. 796–801.

23. Chandorkar, M., Divan, D., Hu, Y., and Banerjee, B. (1994) Novel architectures and control for distributed UPS systems. Applied Power Electronic Conference and Exposition, February 1994, Vol. 2, pp. 683–689.

24. Guerrero, J.M., Vasquez, J.C., Matas, J. *et al.* (2009) Control strategy for flexible micro-grid based on parallel line-interactive UPS systems. *IEEE Transactions on Industrial Electronics*, **26** (3), 726–736.

25. Guerrero, J.M., de Vicuña, L.G., Matas, J. *et al.* (2004) A wireless controller to enhance dynamic performance of parallel inverters in distributed generation systems. *IEEE Transactions on Power Electronics*, **19** (5), 1205–1213.

26. Marwali, M.N., Jung, J.W. and Keyhani, A. (2004) Control of distributed generation systems – Part II: Load sharing control. *IEEE Transactions on Power Electronics*, **19** (6), 1551–1561.

27. Guerrero, J.M., de Vicuña, L.G., Matas, J. *et al.* (2005) Output impedance design of parallel-connected UPS inverters with wireless load-sharing control. *IEEE Transactions on Industrial Electronics*, **52** (4), 1126–1135.

28. Guerrero, J.M., Matas, J., de Vicuña, L.G. *et al.* (2006) Wireless-control strategy for parallel operation of distributed generation inverters. *IEEE Transactions on Industrial Electronics*, **53** (5), 1461–1470.

29. Guerrero, J.M., Matas, J., de Vicuña, L.G. *et al.* (2007) Decentralized control for parallel operation of distributed generation inverters using resistive output impedance. *IEEE Transactions on Industrial Electronics*, **54** (2), 994–1004.

30. Mohamed, Y.A.-R.I. and El-Saadany, E.F. (2008) Adaptive decentralized droop controller to preserve power sharing stability of paralleled inverters in distributed generation microgrids. *IEEE Transactions on Power Electronics*, **23** (6), 2806–2816.

31. Vasquez, J.C., Guerrero, J.M., Liserre, M. and Mastromauro, A. (2009) Voltage support provided by a droop-controlled multifunctional inverter. *IEEE Transactions on Industrial Electronics*, **56** (11), 4510–4519.

32. Guerrero, J.M., Vasquez, J.C., Matas, J. *et al.* (2011) Hierarchical control of droop-controlled AC and DC microgrids – a general approach toward standardization. *IEEE Transactions on Industrial Electronics*, **58** (1), 158–172.

33. Kim, J., Guerrero, J.M., Rodriguez, P. *et al.* (2011) Mode adaptive droop control with virtual output impedances for an inverter-based flexible AC microgrid. *IEEE Transactions on Industrial Electronics*, **26** (3), 689–701.

34. Coelho, E.A., Cortizo, P.C., and Garcia, P.F. (1999) Small signal stability for single phase inverter connected to stiff AC system. Conference Records of the 34th IEEE IAS Annual Meeting, Vol. 4, pp. 2180–2187.

35. Vasquez, J.C., Guerrero, J.M., Luna, A. *et al.* (2009) Adaptive droop control applied to voltage-source inverters operating in grid-connected and islanded modes. *IEEE Transactions on Industrial Electronics*, **56** (10), 4088–4096.

36. Vandoorn, T.L., De Kooning, J.D.M., Meersman, B. *et al.* (2011) Voltage-based control of a smart transformer in a microgrid. *IEEE Transactions on Industrial Electronics*, **26**, 703–713.

37. Savaghebi, M., Jalilian, A., Vasquez, J.C. and Guerrero, J.M. (2012) Autonomous voltage unbalance compensation in an islanded droop-controlled microgrid. *IEEE Transactions on Industrial Electronics*, **60**, 1271–1280.

38. He, J., Li, Y.W. and Munir, M.S. (2012) A flexible harmonic control approach through voltage-controlled DG–grid interfacing converters. *IEEE Transactions on Industrial Electronics*, **59** (1), 444–455.

39. Sharkh, S.M. and Abu-Sara, M. (2004) Digital current control of utility connected two-level and three-level PWM voltage source inverters. *European Power Electronic Journal*, **14** (4), 13–18.

40. Abusara, M.A. and Sharkh, S.M. (2011) Digital control of a three-phase grid connected inverter. *International Journal on Power Electronics*, **3** (3), 299–319.

41. Abusara, M.A. and Sharkh, S.M. (2010) Design of a Robust digital current controller for a grid connected interleaved inverter. Proceedings of the IEEE ISIE, pp. 2903–2908.

42. IEEE (2003) IEEE Standard 1547. IEEE Standard for Interconnecting Distributed Resources with Electric Power Systems, IEEE Press.

10

Microgrid Protection

10.1 Introduction

The integration of distributed generation (DG) within microgrids into distribution networks (DNs) requires rethinking of traditional protection practices to meet new challenges arising from changes in system parameters. For example, fault current's magnitude and its direction can change when DG is introduced into a DN [1,2]. Coordination problems between different protective equipments, that is, between relay, autorecloser, and sectionalizer, can also occur. The level of penetration of DG and the type of interfacing scheme, that is, whether the DG system is based on direct coupling of rotating machines, like synchronous or induction generators, or it is interfaced through a power electronic converter, have a fundamental impact on the protection scheme as these determine the level of short circuit current in the system. The key protection issues which the protection engineer has to address in the new scenario are short circuit power and FCL, device discrimination, reduction in reach of overcurrent (OC) and impedance relays, bi-directionality and voltage profile, sympathetic tripping, islanding, and maloperation of autoreclosers [2–8].

This chapter critically discusses these issues and their solutions in the light of long established rules for the design and coordination of protective relays in a power system, that is, selectivity, redundancy, grading, security, and dependability. Section 10.2 outlines key protection challenges for DNs with DG. Possible solutions to these challenges, including those for an islanded microgrid with inverter interfaced distributed generation (IIDG) units are discussed in Section 10.3. Section 10.4 describes a case study of a typical DN with DG. Finally, Section 10.5 concludes the findings of the chapter.

10.2 Key Protection Challenges

10.2.1 Fault Current Level Modification

Connection to the DN of a single large DG unit or a large number of small DG units, that use synchronous or induction generators, will alter the FCL as both

Power Electronic Converters for Microgrids, First Edition. Suleiman M. Sharkh, Mohammad A. Abusara,
Georgios I. Orfanoudakis and Babar Hussain.
© 2014 John Wiley & Sons, Ltd. Published 2014 by John Wiley & Sons, Ltd.
Companion Website: www.wiley.com/go/sharkh

types of generator contribute to it. This change in FCL can disturb fuse-breaker coordination [2]. A different scenario results when IIDG units are connected to the DN; due to the controller of these interfaces (i.e., inverters), the fault current is limited electronically to typically twice the load current or even less [2–8]. Hence, an independent relay will not be able to distinguish between normal operation and a fault condition without communicating with the inverter. This is true especially in the case of large PV (photovoltaic) installations in a DN, where there is hardly any increase in phase current in the case of a fault or failure. As the fault current is not clearly distinguishable from the operational current, some of the OC relays will not trip; others that might respond would take many seconds instead of responding in a fraction of a second. The undetected fault situation can lead to high voltages despite the low fault currents. Moreover, if the fault remains undetected for long, it can spread out in the system and cause damage to equipment.

The fault impedance can also decrease when DG is introduced into the network in parallel with other devices. The reduced impedance results in high fault levels if the DG unit is a rotating machine or a converter with a low output impedance (e.g., with an output LC filter) without means of isolation from the DN. In the case of a failure, there can be unexpected high fault currents that would put the system components at risk.

The position of a fault point relative to a DG unit and a substation transformer also affects the operation of the protection system. When a fault occurs downstream of the point of common coupling (PCC), both the main source and the DG unit will contribute to the fault current, as shown in Figure 10.1. However, the relay situated upstream of the DG unit will only measure the fault current supplied by the upstream source. As this is only one part of the actual fault current, the relays, especially those with inverse time characteristics, may not function properly, resulting in coordination problems. When the fault is between the main source and a DG unit, then the fault current from the main source would not change significantly as, generally, a DG unit is comparatively small. So in respect of short circuit faults, the incorporation of DG affects the amplitude, direction, and duration of fault currents. The last phenomenon happens indirectly due to the inverse time–current characteristics of relays.

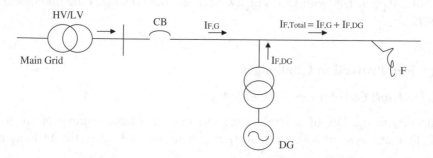

Figure 10.1   Fault current contributions from grid and DG

10.2.2   Device Discrimination

In traditional power systems that have a generation source at one end of the network, as the distance of the fault point from the source increases, the fault current decreases [9]. This is due to the increase in the impedance being proportional to the distance from the source. This phenomenon is used for discrimination of devices that use fault current magnitude. But in the case of an islanded microgrid with IIDG units, as the maximum fault current is limited, so the fault level at locations along the feeder will be almost constant. Hence, the traditional current-based discrimination strategies would not work. New device discrimination strategies will be therefore required to protect the system effectively.

10.2.3   Reduction in Reach of Impedance Relays

The reach of an impedance relay is the maximum fault distance that causes a relay to trigger in a certain impedance zone, or in a certain amount of time. This maximum distance corresponds to a maximum fault impedance or a minimum fault current that is detected [3]. In the case of a fault that occurs downstream of the bus where DG is connected to the utility network, the impedance measured by an upstream relay will be higher than the real fault impedance (as seen from the relay). This is equivalent to an apparently increased fault distance and is due to increased voltage resulting from an additional infeed at the common bus. As a consequence, this will affect the grading of relays and will cause delayed triggering or no triggering at all, as shown in Figure 10.2. This phenomenon is defined as under-reaching of a relay.

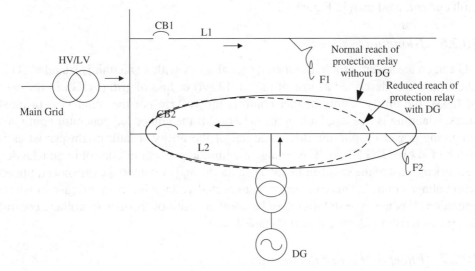

Figure 10.2   Under-reaching of relay and sympathetic tripping fault scenario caused by DG connection

10.2.4 Bidirectionality and Voltage Profile Change

The power flow changes its direction in the case of DNs with embedded DG when local generation exceeds the local consumption [1, 2]. The reverse power flow may hinder the working of directional relays as, traditionally, radial DNs are designed for unidirectional power flow. Moreover, reverse power flow also means a reverse voltage gradient along a radial feeder. This can cause power quality problems, result in violation of voltage limits, and cause increased equipment voltage stress. However, in the case of highly loaded or weak networks, DG can have a positive impact and can improve the network power quality.

DG can also have an impact on the role of tap changing transformers for voltage regulation in DNs. If the location of the DG is close to the network infeed, it affects the tap changing by reducing the load for the transformer. As a result of a shift in tap changing characteristics, the regulation of infeed voltage will be incorrect [2, 3]. Moreover, transformer configuration and grounding arrangements selected for DG connection to the grid must be compatible with the grid, to save the system from voltage swells and overvoltages, and consequent damage [2].

10.2.5 Sympathetic Tripping

This phenomenon can occur due to unnecessary operation of a protective device for faults in an outside zone, that is, a zone that is outside its jurisdiction of operation [3–8]. An unexpected contribution from DG can lead to a situation when a bidirectional relay operates along with another relay which actually sees the fault, thus resulting in malfunctioning of the protection scheme. For example, a relay at line L2 can unnecessarily operate for a fault F1 at line L1 as a result of infeed from DG to the fault current, as shown in Figure 10.2.

10.2.6 Islanding

DG can create severe problems when a part of a DN with a DG unit is islanded. This phenomenon is described as loss of mains (LOM) or loss of grid (LOG). In the case of LOM, the utility supply neither controls the voltage nor the frequency. In most cases, islanding is due to a fault in the network. If the embedded generator continues supplying power despite the disconnection of the utility, a fault might persist as it will be fed by a DG [2–8]. The voltage magnitude gets out of control in an islanded network as most of the small embedded generators and grid interfaces are not equipped with voltage control. This can lead to unexpected voltage levels in the case of island operation. Frequency instability may be another result of the lack of voltage control that poses a risk to electric machines and drives.

10.2.7 Effect on Feeder Reclosure

The role of an autorecloser is very important in restoring the system after a fault that lasts for a very short interval. However, in the case of a DN with DG, two main problems may result from automatic reconnection of the utility after a short interval [2–8]. The first problem is that the automatic recloser attempt may fail as a result of feeding of a fault from a DG. The second problem is that due to active power imbalance, a change in frequency may occur in the islanded part of the grid. In this scenario, an attempt at reclosing the switch would couple two asynchronously operating systems. Moreover, conventional reclosers are designed to reconnect the circuit only if the substation side is energized and the opposite side is unenergized. However, in the case of DG, there would be active sources on both sides of the recloser, thus hampering its working.

10.3 Possible Solutions to Key Protection Challenges

There are several possible solutions to cope with the new challenges caused by introduction of DG into DNs. To solve the problem of bidirectionality, the main relays of the feeders which are fed from the same substation can be interlocked [6]. The main relay of the feeder with DG will be equipped with the interlocking system. Once a short circuit is detected by a relay, it will send a locking signal to the main relay of the feeder having DG. Due to this locking signal, the main relay will not maloperate, even if there is back feeding from the DG to the fault. The use of directional OC relays instead of OC relays can also solve this problem, but this scheme has it own limitations, as mentioned earlier. Main feeder's relays readjustment in terms of time settings is another solution. The feeder without DG can have faster relay settings than the relay settings of the feeder with DG [6]. But care has to be taken with this readjustment so that it does not hinder the coordination of these relays with downstream protection devices of the feeder.

Generally, disconnection of the DG from the network (by means of interconnect protection) is ensured before reclosing the feeder breaker. Use of a communication channel between a substation and the DG to transfer trip a DG unit can ensure fast reconnection. In cases where DG is allowed to carry islanded loads, a syn-check relay on the circuit breaker or recloser that coordinates synchronization with the grid can be used to reclose the feeder breaker [8].

There is a trade-off between the speed of reclosing and the power quality, that is, the faster the reclosing, the better the power quality. However, to ensure that the reclosing attempt is successful, instantaneous reclosing is not recommended for feeders with DG. Increase of the reclose interval from the usual 0.3 s to 1 s is recommended for such feeders [3, 8].

10.3.1 Possible Solutions to Key Protection Challenges for an Islanded Microgrid Having IIDG Units

Conventional protection schemes face serious challenges when it comes to protecting an islanded microgrid with IIDG units. They need major revision to find new methods based on the limited fault current to detect and isolate the faulty portion. The various possible solutions to cope with the problem can be broadly divided into four categories: use of inverters having high fault current capability, that is, uprating of the inverter [10]; communication between the inverter and protective relays; introduction of energy storage devices that are capable of supplying large current in case of a fault [11]; and in-depth analysis of the fault behavior of an islanded microgrid with an IIDG unit to comprehend the behavior of system voltages and currents [12, 13]. This will in turn help in defining alternative fault detection and alternative protection strategies that, in case of a fault, do not rely on a large magnitude of the fault current but rely on other parameters, like change in the voltage of the system [14].

10.3.1.1   Differential Protection Schemes

In [15], the authors have proposed the application of a differential protection scheme, which is traditionally used for transformer protection, for the protection of an islanded microgrid. This scheme based on differential relays is selected as its operation, unlike OC relays, is independent of the fault current magnitude. This scheme solves the problem of low fault current in the case of IIDG, but it has a downside, too. The protection scheme would not be able to differentiate between a fault current and an overload current, so nuisance tripping will result whenever the system is overloaded. So, traditional differential protection schemes might not be, in some instances, able to differentiate correctly between internal faults and other abnormal conditions. Also mismatch of the current transformers can be a source of malfunction.

For a coordinated clearing of a fault in an islanded microgrid and to ensure selectivity, it is important that different distributed generators can communicate effectively with each other. To this end, evolving a distribution system version of the pilot wire line differential protection may be needed [7].

10.3.1.2   A Balanced Combination of Different Types of DG Units for Grid Connection

Another way to ensure the proper protection of an isolated microgrid is to use DG units with synchronous generators, or to use inverters having high fault current capability, or to use a combination of both types of DG unit, so that conventional protection schemes can be properly used. This combination will ensure a large fault current that can be detected by conventional protection schemes. However, for a higher rated inverter, large size power electronic switches, inductors, and capacitors, etc, would be needed, thus making the system expensive. Energy storage devices, like batteries and flywheels, can also be incorporated into the microgrids to increase the fault level to a desired level. In the case of low voltage circuits, the fault current should be at least three times greater than the maximum load current for its clearance by OC relays [7].

Directional relays can be used to clear the fault within the microgrid provided they see a fault current exceeding the maximum load current in their tripping direction. But this is not always the case as faults can be fed from different directions, as explained in Section 10.2.5.

10.3.1.3 Inverter Controller Design

A protection scheme for an islanded microgrid is heavily dependent on the type of the inverter controller as the controller actively limits the available fault current from an IIDG unit. This has been demonstrated in [13, 16] where two different controllers, that is, one using "dq0" coordinates and the other using three-phase (*abc*) coordinates, are employed to control a stand-alone four leg inverter supplying a microgrid. In both cases, the fault current is quite small but its magnitude is different. Thus, selection of a controller can be important for microgrid protection.

10.3.1.4 Voltage-Based Detection Techniques

A protection scheme that combines conventional OC characteristics and undervoltage initiated directional fault detection with definite time delays is proposed in [9]. A large depression in network voltage cannot be used alone for detecting low levels of fault current in a microgrid as voltage depression would not have sufficient gradient to discriminate the protection devices. Hence, measurement of some other parameters is recommended. It is mentioned in [9] that simple device discrimination can be achieved by current direction together with definite time delays. The duration of delays is proposed to be set on the basis of sensitivity of loads or generation to under voltage. For setting up adequate discrimination paths, selecting different delays for the forward and reverse direction flows of the fault current is recommended. This scheme looks sound but the use of communication channels for coordinating protection with control and automation schemes can complicate things.

The authors of [14] propose various voltage detection methods to protect networks with a low fault current. One of the suggested methods makes use of the Clarke and Park transformations [17] to transform a set of instantaneous three-phase utility voltages into a synchronously rotating two-axis coordinate system. The resultant voltage is compared with a reference value to detect the presence of the disturbance. In the case of an unsymmetrical fault, the utility voltage "dq" components have a ripple on top of the DC term. Hence, these components are first notch filtered and then compared with the reference. Fault detection in the case of low fault current networks can be achieved by making use of voltage source components. It is possible to calculate the values of voltage source components for different types of faults since the theory of the interconnection of equivalent sequence networks in the event of a fault is applicable here [16].

Another voltage detection method makes use of the fact that the sum of two squared orthogonal sinusoidal waveforms is equal to a constant value [18]. The necessary 90° phase shift is achieved by passing each phase voltage through an all-pass filter. The output is obtained by summing the squared values of the two signals for each phase and then comparing them with the reference after filtration to detect any disturbance.

The above-mentioned schemes have their limitations. The performance of the schemes can suffer due to time delays and filtering actions. The time required for detection of a fault in each case is different as it depends upon the type of the fault as well as on the magnitude of the voltage at the faulty feeder at the moment of the occurrence of the fault. Time delay is also introduced by the filtering action.

10.3.1.5 Adaptive Protection Schemes

Adaptive protection schemes are presented as a solution for microgrid protection both in grid-connected and in islanded mode of operation. The basic philosophy behind these schemes is automatic readjustment of the relay settings when the microgrid changes from grid-connected to islanded mode and vice versa. In an islanded microgrid, the adaptive protection strategy can be used by assigning different trip settings for different levels of fault current, which in turn are linked to different magnitudes of system voltage drops resulting from disturbance in the system.

As discussed earlier, in an islanded microgrid with IIDG units, the fault in the system can result in severe voltage depression in the entire network (due to low impedances within the network). In such a scenario, selectivity cannot be assured using voltage measurement alone. A possible solution could be the use of a voltage restrained OC technique, as proposed in [14]. The scheme is shown in Figure 10.3. A large depression in voltage (which happens mostly in the case of short circuit

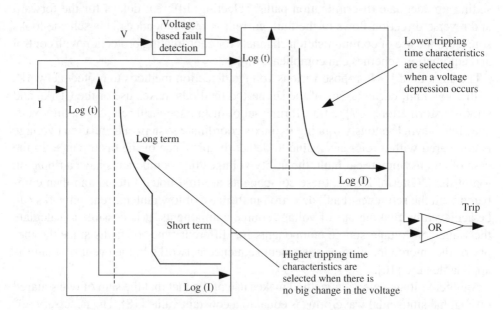

Figure 10.3   Voltage restrained overcurrent protection scheme logic circuit [14]

as opposed to overload) will result in the selection of a lower current threshold. This would effectively move the time–current characteristic down and thus the tripping time would be reduced. In contrast to this, tripping times would be longer during overloads as a small voltage depression would not be able to switch the scheme to the lower setting. Thus the system would retain the longer time setting corresponding to long-term characteristics.

This scheme, although looks sound, has its drawbacks. It is not clear how the scheme will perform if there is a small difference in the magnitude of voltage depression resulting from short circuit and overload conditions. The scheme seems to make use of the principle of relays with inverse time characteristics, that is, the larger the fault current, the smaller the response time. Also, it will suffer from a long clearing time in the case of faults that cause a small voltage drop, thus posing the risk of fault current spreading in the entire network.

See also Chapter 11, which discusses an adaptive fuse saving protection scheme for a grid with DG.

10.3.1.6 Protection Based on Symmetrical and Differential Current Components

An islanded microgrid can be protected against single line-to-ground (SLG) and line-to-line (LL) faults with a protection strategy that makes use of symmetrical current components [19]. Based on these facts, a symmetric approach for protection of a microgrid is proposed in [20]. This scheme makes use of differential and zero-sequence current components as a primary protection for SLG faults and negative-sequence current components as a primary protection against LL faults. It is recommended that a threshold should be assigned to each of the symmetrical current components to prevent the microgrid protection from operating under imbalanced load conditions. This threshold should be selected carefully to avoid any maloperation of relays.

10.4 Case Study

Figure 10.4 shows a single line diagram of the system that is simulated to investigate the impact of DG on DN protection. A typical 25 kV DN is configured downstream of a 69/25 kV substation named as the main substation (MS). The utility grid upstream of the substation is represented by a Thevenin equivalent of voltage source and series impedance with short circuit level of 637 MVA and an X/R ratio of 8 at 69 kV bus. The MS is equipped with a 69/25 kV, 15 MVA load tap-changing transformer, with delta-wye grounded configuration. The transformer has a series equivalent impedance of 7.8% at 15 MVA base and connects the DN to the 69 kV sub-transmission system.

The DN is modeled by two load feeders, LF1 and LF2 in Figure 10.4, emanating from the 25 kV bus. The system load – 10 MW on each feeder – is modeled as a

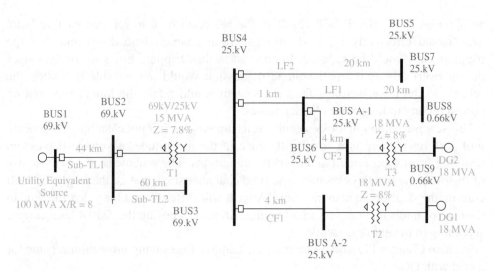

Figure 10.4   Single line diagram of a typical distribution network (DN) with distributed generation (DG) where sub-TL stands for sub-transmission line, CF and LF represent collector feeder and load feeder, respectively

constant impedance that has no contribution to the fault. Two equivalent synchronous generators rated at 18 MVA are connected to the 25 kV bus through two 0.66/25 kV, 18 MVA step-up transformers – with delta-wye grounded configuration – and through two 25 kV collector feeders, CF1 and CF2 in Figure 10.4.

The positive sequence impedance for 25 kV feeders is $0.2138 + j0.2880\,\Omega\,km^{-1}$ and for a 69 kV feeder it is $0.2767 + j0.5673\,\Omega\,km^{-1}$. The zero sequence impedance for the 69 kV sub-transmission line (Sub-TL) is $0.5509 + j1.4514$.

A distance relay (SEL 321) is installed at the bus 1 end of the Sub-TL1 to protect against faults at Sub-TL1 and Sub-TL2 and to provide back-up protection for some part of the DN. An OC fuse rated at 200 A is installed on the high voltage side of transformer T1 to provide protection against transformer internal faults and back-up for feeder faults. The load feeders are equipped with time-graded OC and earth fault (EF) relays (i.e., 51/51 N) and instantaneous OC and EF relays (i.e., 50/50 N) for protection against phase and ground faults. The collector feeders are also protected by OC relays. OC and the EF relays of load feeders LF1 and LF2 are set at 280 and 140 A respectively. The OC relays at both the collector feeders are set at 400 A.

ASPEN OneLiner was used to simulate different faults for determining short circuit levels and to investigate their impact on protection coordination, including reach of distance relays. Figure 10.5 shows the time–current characteristic curve of the fuse installed at the high voltage side of the MS transformer and the OC relays installed at the load feeders. The operating times that are shown on the curve are of the fuse and OC relays of the load feeders in the case of a three-phase to ground (3LG) fault at 90 % of the LF1 length.

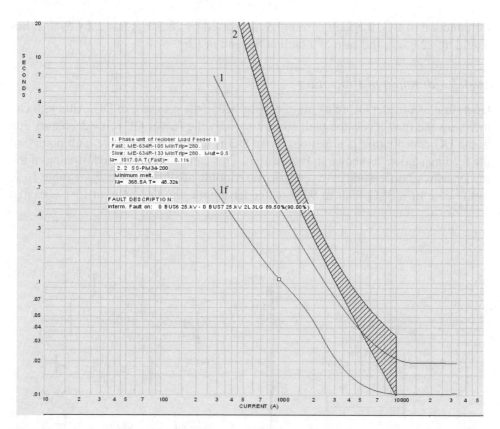

Figure 10.5 Operating times of OC relay installed at LF1 and fuse located at high voltage side of substation transformer for 3LG fault at 90% of LF1 length without DG. '1f' and '1' stand for the recloser's fast and slow characteristics curves respectively while '2' represents the fuse characteristics curve. The relay current and operating time are shown both as points on the curves and as text within the description boxes.

10.4.1 Fault Level Modification

A 3LG fault was applied to determine the fault current at different points with and without DG connection, as shown in Table 10.1. It is clear from the table that after

Table 10.1 Fault currents at different network buses with and without DG

Fault current (A)					
Without DG while three phase to ground fault is at			With DG while three phase to ground fault is at		
Bus 2	Bus 4	End of LF1	Bus 2	Bus 4	End of LF1
1147	1874	967	1605	3826	1252

introduction of DG, the fault current has increased by 28.5% at bus 2, by 51% at bus 4, and by 22.8% in the case of a fault at the end of LF1.

10.4.2   Blinding of Protection

Operation of a feeder OC relay may be disturbed in the presence of DG. Although DG increases the fault levels, the fault current seen by the feeder OC relay decreases due to the DG contribution in situations when DG is located between the fault point and the feeding station, as shown in Figure 10.6. This can result in delayed tripping of the feeder relay or, in a worst case scenario, no tripping at all. It is clear from Table 10.2, in the case of a 3LG fault at 90% of feeder length, the OC relay at LF1 operated

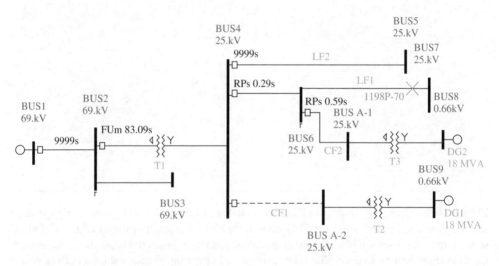

Figure 10.6   Blinding of protection or delayed tripping scenario in the case of a 3LG fault at 90% of LF1 length with DG2 connection only

Table 10.2   Operating times of protection devices in the case of a 3LG fault at 90% of the LF1 length (N/O stands for no operation and DR stands for distance relay)

Configuration of DN	Operating time (s)				
	DR (Zone 3)	Fuse	OC relay at LF1	OC relay at CF 1	OC relay at CF 2
Without DG	N/O	48.32	0.23	N/O	N/O
With DG 1	N/O	79.85	0.18	0.58	N/O
With DG 2	N/O	82.09	0.29	N/O	0.59
With both DGs	N/O	64	0.29	0.89	0.9

in 0.23 s when no DG was connected and the same relay operated in 0.29 s when only DG2 was connected or when both DG units were connected.

10.4.3 Sympathetic Tripping

Sometimes DG can contribute to a fault on a feeder fed from the same substation or even to a fault at higher voltage levels, resulting in unnecessary isolation of a healthy feeder or a DG unit. For example, an OC relay at CF1 can unnecessarily operate for a high resistive 3LG fault at LF1 as a result of infeed to the fault from DG1 through the substation bus bar, as shown in Figure 10.7.

10.4.4 Reduction in Reach of Distance Relay

Distance relays are set to operate in a specific time for any faults occurring within a predefined zone of a transmission line or a distribution feeder. Due to the presence of DG, a distance relay may not operate according to its defined zone settings. When a fault occurs downstream of the bus where DG is connected to the utility, impedance measured by an upstream relay will be higher than the real fault impedance (as seen from the relay). This can disturb the relay zone settings and can, thus, result either in delayed operation or, in some cases, no operation at all.

Table 10.3 shows the zone settings for the distance relay installed at the Sub-TL 1 (shown in Figure 10.4). It is clear from the table that the range of zone 2 decreases to 67% when DG is connected from 79% when DG was not connected. Similarly, the

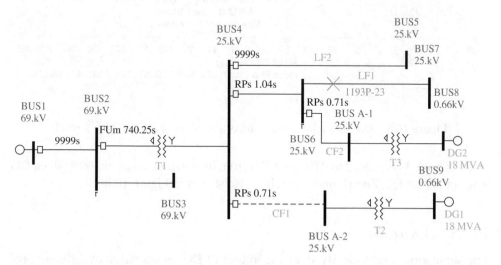

Figure 10.7 Sympathetic tripping scenario when relay at CF1 opens for a high resistive 3LG fault at 30% of LF1 length with both DG connected

Table 10.3   Operating zones of distance relay with and without DG

Zones	Relay settings (% of line length)	Distance relay operating range			
		Three-phase fault		One-phase fault	
		Without DG	With DG	Without DG	With DG
Zone 1	40	40	40	39	39
Zone 2	80	79	67	79	74
Zone 3	115	100	91	100	100

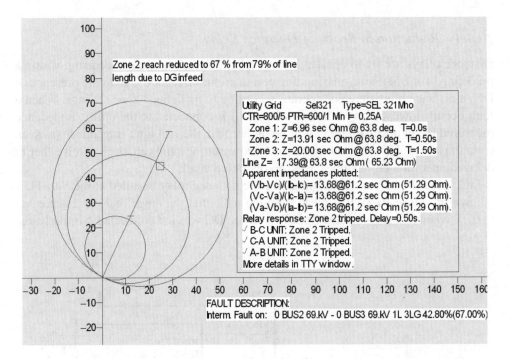

Figure 10.8   Distance relay zone settings and effect of DG on Zone 2 reach

reach of zone 3 is reduced to 91% with DG from its previous value of 100% when DG was not connected. Zone 2 under-reach is also shown in Figure 10.8.

10.4.5   Discussion

The simulation results clearly show the impact of DG on operation of different protective relays. Most of the problems that may arise can be solved by the use of modern microprocessor based multi-function relays with more features and readjustment of relay settings. For example, replacing simple OC relays with directional relays on collector feeders can solve the problem of sympathetic tripping. In this case, additional relays would be required to provide backup protection to 25 kV bus bar. To solve the problem of delayed tripping, current setting of the relay at LF1 can be lowered. But this needs great care, as if the setting is too low, nuisance tripping can occur in severe overload conditions. Another solution is to have relays with two groups of settings; one group for operation without DG and the other group for operation with DG. For its working, this scheme requires a communication link between the OC relays at LF1 and the relays at collector feeders.

So far as under-reaching of the distance relay is concerned, readjustment of zone settings or addition of an extra zone can make the relay to operate in the correct zone. However, the distance relay can over reach if the DG units are disconnected. Therefore, these conditions should be checked to ensure that there is no miscoordination with the adjacent Zone 2. Installation of one 25 kV circuit breaker and a multifunction protective relay including 50/51, 50/51N, 67,67N and other functions can provide fast interruption of fault current infeed for transformer faults, primary protection for the 25 kV bus bars and local backup protection in case of failure of collector feeder relays. The preferred location for this circuit breaker is the LV side of MS transformer.

10.5 Conclusions

Different protection issues caused by integration of DG into DNs have been discussed. By making use of a simulation model of a typical DN, different fault scenarios with and without DG, have been simulated and the behavior of an existing protection set-up has been examined. The results show that DG integration can change the FCL and, consequently, coordination of protection devices. Nuisance tripping of relays can also occur. Distance relays can under-reach as a result of fault current infeed from DG. All these issues should be addressed in order to ensure that protection of the system remains reliable even after introduction of DG. Various protection schemes for microgrids, both in grid-connected and islanded mode of operation, are explored. The techniques and strategies, such as analysis of current using digital signal processing for characteristic signatures of faults [21], development of a real-time fault location technique having the capability to determine the exact fault location in all situations, use of impedance methods, zero sequence current and/or voltage detection-based relaying, and differential methods using current and voltage parameters [22], have the potential for developing more robust protection schemes to cope with the new challenges.

References

1. Jenkins, N., Allan, R., Crossley, P. *et al.* (2000) *Embedded Generation*, IEE Power and Energy Series, vol. 31, IEE Press, London.
2. Barker, P.P. and De Mello, R.W. (2000) Determining the impact of distributed generation on power system: part 1 – radial distribution systems. *IEEE Power Engineering Society*, **3**, 1645–1656.
3. Geidl, M. (2005) Protection of Power Systems with Distributed Generation: State of the Art, Power Systems Laboratory, ETH Zurich.
4. Maki, K. S. Repo, and P. Jarventausta (2004) Effect of wind power based distributed generation on protection of distribution network. 8th International Conference on Developments in Power System Protection, pp. 327–330.
5. EPRI (2005) Distribution System Design for Strategic Use of Distributed Generation, EPRI, Palo Alto, CA.
6. Abdel-Galil, T.K., Abu-Elnien, A.E.B., El-saadany, E.F., *et al.* (2007) Protection Coordination Planning With Distributed Generation. Report Number -CETC 2007-49. Qualsys Engco. Inc..
7. Feero, W.E., Dawson, D.C., and Stevens, J. (2002) White paper on Protection Issues of The MicroGrid Concept, CERTS.
8. Power System Relaying Committee, IEEE PES (2004) The Impact of Distributed Resources on Distribution Relaying Protection.
9. Tumilty, R.M., Elders, I.M., Burt, G.M. and Mcdonald, J.R. (2007) Coordinated protection, control and automation schemes for microgrids. *International Journal of Distributed Energy Resources*, **3**, 225–241.
10. Lopes, J.A.P., Moreira, C.L. and Madureira, A.G. (2006) Defining control strategies for microgrids Islanded operation. *IEEE Transactions on Power Systems*, **21** (2), 916–924.
11. Jayawarna, N., Jenkins, N., Barnes, M., *et al.* (2005) Safety analysis of a microgrid. International Conference on Future Power Systems, pp. 1–7.
12. Timbus, A.V., Rodriguez, P., Teodorescu, R. *et al.* (2006) Control strategies for distributed power generation systems operating on faulty grid. IEEE International Symposium on Industrial Electronics, Vol. 2, pp. 1601–1607.
13. Baran, M.E. and El-Markaby, I. (2005) Fault analysis on distribution feeders with distributed generators. *IEEE Transactions on Power Systems*, **20** (4), 1757–1764.
14. Tumilty, R.M., Brucoli, M., Burt, G.M. and Green, T.C. (2006) Approaches to network protection for inverter dominated electrical distribution systems. *Power Electronics, Machines and Drives (PEMD)*, **1**, 622–626.
15. Zeineldin, H.H., El-Saadany, E.F., and Salama, M.M.A. (2006) Distributed generation micro-grid operation: control and protection. Power Systems Conference, pp. 105–111.
16. Brucoli, M. and Green, T.C. (2007) Fault behaviour in islanded microgrids. 9th International Conference on Electricity Distribution, Vienna, Austria, pp. 21–24.
17. Sannino, A. (2001) Static transfer switch: analysis of switching conditions and actual transfer time. IEEE Power Engineering Society Winter Meeting, 2001, Vol. 1, pp. 120–125.
18. Ise, T., Takami, M., and Tsuji, K. (2000) Hybrid transfer switch with fault current limiting function. Proceedings of the 9th Int. Conference on Harmonics and Quality of Power, Vol. 1, pp.189–192.

19. Bergen, A.R. (1986) *Power System Analysis*, Prentice Hall Publications.
20. Nikkhajoei, H. and Lasseter, R.H. (2006) Microgrid Fault Protection Based on Symmetrical and Differential Current Components, CEC, Public Interest Energy Research Program.
21. Kueck, J.D. and Kirby, B.J. (2003) The distribution system of the future. *The Electricity Journal*, **16** (5), 78–87.
22. Lasseter, R.H., Akhil, A.A., Marnay, C. *et al.* (2002) White Paper on Integration of Distributed Energy Resources: The CERTS MicroGrid Concept, CERTS.

11

An Adaptive Relaying Scheme for Fuse Saving

11.1 Introduction

The basic objective of protection coordination in power systems is to achieve selectivity, that is, to switch off only the faulty component and to leave the rest of the power system in service in order to minimize discontinuity of supply and to ensure stability. Proper coordination ensures that there is neither a maloperation of protective devices nor a duplication of their operation. The traditional coordination between a circuit breaker (CB), a recloser, and a lateral fuse in a typical distribution network (DN) is illustrated in Figure 11.1 [1]. To achieve selectivity in the isolation of a fault, protective devices in series are time coordinated such that the upstream device electrically closest to the short circuit or overload opens first to isolate the faulty section. However, to ensure reliability of supply, that is, to avoid prolonged discontinuity of service due to temporary faults, utilities normally employ a fuse saving strategy. Through this strategy, utilities try to "save" the fuse on the circuit following temporary faults by de-energizing the line with the fast operation of an upstream interrupting device (i.e., a recloser) before the fuse has a chance to blow. The interrupting device then recloses and restores power when the fault is cleared, and the fuse is saved. This is a useful practice as it saves expensive fuse replacement and unnecessary extended service interruption which otherwise could be quite burdensome as temporary faults constitute almost 70–80% of faults occurring in DNs. In the case of a permanent fault, upon completion of recloser fast mode shots, the fuse blows to clear the fault while the recloser is waiting to operate on its slow curve. The feeder head-end CB provides an overall back-up protection since its time–current characteristics (TCC) curve is above all the other curves.

As shown in Figure 11.1, the coordination between a recloser and a fuse will be valid only if the fault current is between the minimum fault current of the feeder ($I_{F\text{-min}}$)

Power Electronic Converters for Microgrids, First Edition. Suleiman M. Sharkh, Mohammad A. Abusara, Georgios I. Orfanoudakis and Babar Hussain.
© 2014 John Wiley & Sons, Ltd. Published 2014 by John Wiley & Sons, Ltd.
Companion Website: www.wiley.com/go/sharkh

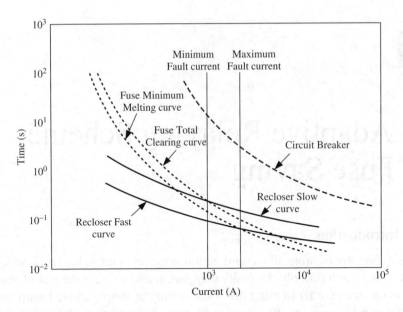

Figure 11.1   A typical time–current coordination (TCC) curve for protective devices

and the maximum fault current of the feeder ($I_{\text{F-max}}$), and provided that there is some margin between the operating times of these two devices. This is because between $I_{\text{F-min}}$ and $I_{\text{F-max}}$, the recloser's fast characteristic curve lies below the minimum melt (MM) characteristic of the fuse and the recloser's slow characteristic curve lies above the total clearing (TC) characteristic of the fuse. However, integration of distributed generation into DNs, depending upon the size, type, and location of the DG, may affect the recloser and fuse coordination (RFC) [1–3]. Recloser–fuse (RF) miscoordination can occur when the fault current becomes greater than $I_{\text{F-max}}$ due to the additional fault current contribution from the DG. It can also happen when, due to the location of the DG, the fault current seen by a recloser is less than the current that passes through a fuse, even when the fault current is between $I_{\text{F-min}}$ and $I_{\text{F-max}}$. As a result, the fuse could blow first or both the fuse and the recloser could operate at the same time.

Various measures proposed in the literature for maintaining protection coordination in power systems, particularly for maintaining RFC in DNs with DG, may be broadly divided into two categories: preventive and remedial. Preventive approaches are those which aim at limiting the contribution of DG to such an extent that the original RFC of the network is not disturbed, as proposed in [4–9]. Remedial solutions are those which recommend changes in settings of protective devices to maintain RFC in the presence of DG, as discussed in [2, 10–22].

11.1.1   Preventive Solutions Proposed in the Literature

In [4], the authors recommend that protection selectivity should be checked whenever a new DG unit is connected to a DN and settings of overcurrent (OC) devices should be recalculated to maintain it. In [5–7], a method based on mathematical equations of the characteristics of protective devices is used to determine the maximum size of DG that can be integrated into the system without causing RF miscoordination. In [8] it is suggested that, generally, the only way to maintain RFC without modifying the protection scheme, in the presence of DG, is an instant disconnection of all DG connected to the system, in the case of a fault. Once DG is offline, the original protection coordination, that is, the one prior to DG connection, will be restored. However, this scheme cannot work in conditions where the fault level is such that the coordination margin between operating times of the recloser and fuse is almost negligible. By the time the DG is disconnected, the fuse would already have blown. The use of a fault current limiter (FCL) for maintaining protection coordination when DG is connected to a DN is discussed in [9]. However, the FCL is a relatively new device and needs more practical testing and field experience before becoming a reliable solution for fault current limiting in electric networks. In a DN with a high level of DG penetration, optimal location of an FCL and selection of its best impedance may be challenging.

11.1.2   Remedial Solutions Proposed in the Literature

1. **Miscellaneous:** By showing the impact of different DG types and sizes on the fuse melting time, the authors of [2] propose changing the fuse size, or selecting faster recloser settings with fuse replacement as the most likely option for ensuring RFC. A change in the type or size of a fuse is also proposed in [10] when it is not possible to save it over the entire fault current range. In circumstances where it is not possible to save a fuse due to a high fault contribution from DG, use of a sectionalizer as a solution is discussed in [11].

   Reference [12] proposes modifying the TCC of the recloser fast curve for maximum possible fault current on the basis of the fuse to recloser currents ratio, that is, ratio of the fuse current I_F to the recloser current I_R (I_F/I_R) and then coordinating it with the fuse curve. To this end, the use of an adaptive microprocessor-based recloser is recommended. In [13], it is proposed to modify the TCC of the recloser's fast curve based on the minimum value of I_R/I_F to ensure RFC even in the presence of DG. To accomplish this, the use of a microprocessor-based recloser is recommended. It is shown that the revised characteristic curve is valid even when DG is disconnected. However, the schemes discussed in [12, 13] may not be very effective when it comes to maintaining coordination in the worst fault conditions, as will be discussed later in this chapterer.

   To avoid simultaneous operation of the recloser fast curve and fuse in the case of high fault currents, a partial fuse saving scheme is described in [14]. The fuse characteristic curve is truncated at the maximum coordination current (MCC) which is defined as the current where fuse saving can no longer be guaranteed by the recloser. However, with this scheme, it will be difficult to define the MCC when DG is connected to the network, that is, it will vary depending upon the size and location of the DG.

2. **Adaptive relaying:** References [15–17] discuss strategies relying on an adaptive I_{pickup} of OC relays to detect faults (normal or high impedance) in DNs that otherwise remain undetected due to fixed settings of OC relays or due to the introduction of DG into the network. Most of these strategies rely on modifying relay settings in real time as the operating conditions of the system change, that is, I_{pickup} is updated according to currently available short circuit current and actual loading conditions.

   Adaptive protection schemes described in [18, 19] propose dividing the DN into various zones with a reasonable balance of load and DG in each zone. The schemes aim at maintaining protection coordination while keeping most of the DG online during a fault by allowing islanded operation of the DG. However, due to the location of the DG units with respect to the loads, the fluctuating nature of power from these DG units and uncertainty of utility loads, it might not be possible to establish zones that fulfill the required criteria. Moreover, islanded operation of DG may not be desirable; according to the current utility practice, islanding is not allowed.

   A scheme is proposed in [20] to keep protection coordination intact during a fault without creating any island or DG disconnection while ensuring that the conventional OC protection devices, that is, OC relays, reclosers, or fuses, do not lose their proper coordination, but this scheme is intended only for meshed networks.

3. **Expert systems and agent-based schemes:** Expert systems and multi-agent approaches have been discussed in [21, 22] to solve protection coordination problems in distribution systems. However, these systems are expensive as well as difficult to realize and maintain due to their complexity.

11.1.3   Contributions of the Chapter

This chapter describes an alternative protection strategy that benefits from the characteristics of both time and instantaneous OC (i.e., 51 and 50) elements for maintaining RFC in a typical DN with DG. Moreover, a novel algorithm is proposed to adaptively change the setting of the 50 element to ensure fuse saving under the worst fault conditions. Section 11.2 introduces a case study of a typical DN with DG. In Section 11.3, the potential impacts of DG on RFC are investigated through simulations of different fault scenarios with different DG configurations and the results are analyzed to develop a solution strategy. In Section 11.4, a scheme is proposed to restore proper RFC for fuse saving even in the presence of DG. A reclosing strategy for the feeder with DG is described in Section 11.5. In Section 11.6, some observations are made in respect of the application and implementation of the proposed strategy. Finally, Section 11.7 concludes the chapter.

11.2   Case Study

Figure 11.2 shows a single line diagram of a typical North American DN section (henceforth referred to as the test system), which was described in [23]. This test system is simulated to demonstrate and investigate the impact of DG on a fuse saving

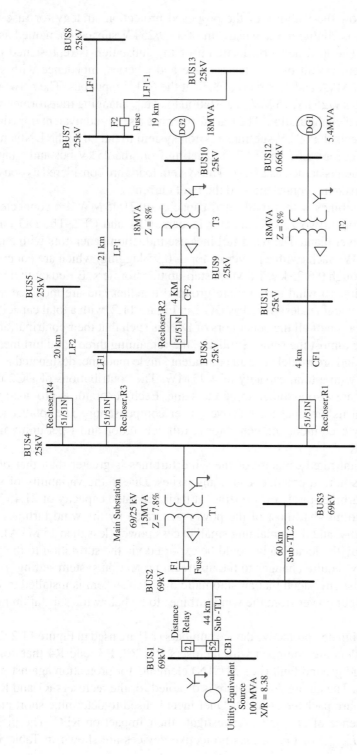

Figure 11.2 Single line diagram of a section of a typical distribution network (DN) with distributed generation (DG) where Sub-TL stands for sub-transmission line, CF and LF represent collector feeder and load feeder, respectively. IEEE/ANSI designated protective device numbers [24] are used

scheme and to show the viability of the proposed protection strategy for fuse saving. A typical DN is configured downstream of a 69/25 kV substation, named as the main substation. The grid network upstream of the substation is represented by a Thevenin equivalent circuit of a voltage source and a series impedance with short circuit level of 637 MVA and X/R ratio of 8.38 at the 69 kV input bus. The network is equipped with a 69 kV/25 kV, 15 MVA substation load tap-changing transformer, with delta-wye grounded configuration. The transformer has a series equivalent impedance of 7.8% and connects the 69 kV subtransmission system to the DN. The DN is modeled as two load feeders, LF1 and LF2, emanating from the 25 kV bus and a tapped line from LF1, which is named as LF1-1. The system loads are considered as constant impedances that make no contribution to the fault current.

Thirteen wind turbines with a total rated capacity of 21.45 MW are connected to the 25 kV bus through two 25 kV collector feeders, CF1 and CF2. The DG unit is an actual wind power plant. The wind-turbines use induction generators with a rated capacity of 1.65 MW each (with a power factor of 0.95 lagging) which are connected to the network through 0.6/25 kV, 1.8 MVA step-up transformers. Based on their geographical proximity, 10 wind turbines are grouped together and are modeled as an equivalent single machine, designated as DG 1 in Figure 11.2, with a total capacity of 16.5 MW. The location of all the generators of DG1 is such that their contributions to different faults are almost the same. Similarly, the remaining three wind turbines are grouped together and are modeled as an equivalent single machine, designated as DG 2 in Figure 11.2, with a total capacity of 4.95 MW. The contributions of DG2 units to different faults are also assumed to be the same. Each generator is provided with capacitor banks at its terminal for reactive power compensation. The total reactive power required by each induction generator at full generation and 1 per unit voltage is about 740 kVAR.

Although the total rated capacity of the wind turbines is greater than that of the substation, it does not, in practice, cause any issues. Due to the variability of wind speed, the wind turbines rarely ever operate at their full rated capacity of 21.45 MW. It is also worth noting that most of the power generated by the wind turbines will be consumed locally, and the remaining small excess power (less than 15 MVA), that is not consumed by the local loads, could be exported via the substation to the grid; this of course may require changes to the upstream protection system setting. Alternatively, it could be envisaged that an automatic dispatch system is installed to limit the maximum output power from the wind turbines to be below the substation rating of 15 MVA.

IEEE/ANSI designated protective device numbers [24] are used in Figure 11.2. Load and collector feeders are equipped with reclosers R1, R2, R3, and R4 that contain time OC phase and ground fault (i.e., 51/51N) elements for protection against phase and ground faults. The autoreclose feature is disabled for the reclosers R1 and R2. A commercial software package (ASPEN OneLiner) is used to determine short circuit levels in the presence of DG and to investigate their impact on RFC. The pick-up current settings (I_{pickup}) of the various protective devices are shown in Table 11.1.

Table 11.1 Current settings of protection devices installed in the test system shown in Figure 11.2

Protection device	OC relay at LF1	OC relay at LF2	Earth fault relay at LF1	Earth fault relay at LF2	OC relay at CF1	OC relay at CF1	Fuse F2
Settings (A)	290	290	140	140	450	450	380

The impedance values of the sub-transmission lines (Sub-TLs) and the feeders are as follows:

$$Z^+{}_{25\,kV\ feeders} = 0.2138 + j0.2880\,\Omega\,km^{-1}$$
$$Z^+{}_{69\,kV\ Sub\text{-}TL} = 0.2767 + j0.5673\,\Omega\,km^{-1}$$
$$Z^0{}_{69\,kV\ Sub\text{-}TL} = 0.5509 + j1.4514\,\Omega\,km^{-1}$$
$$B^+{}_{Sub\text{-}TL} = 0.00803014065\,\Omega\,km^{-1}$$
$$B^0{}_{Sub\text{-}TL} = 0.00481803678\,\Omega\,km^{-1}$$
B stands for line charging.

11.3 Simulation Results and Discussion

To illustrate the impact of DG on the fuse saving scheme, the time coordination between the recloser R3 and the fuse F2 (150 K) of the test system shown in Figure 11.2 has been investigated. The coordination is represented by the TCC of the recloser and MM and TC characteristics of the fuse in Figure 11.3 (mathematical description of the TCC is given in [6]). Two reclosing attempts, that is, one in fast mode and one in slow mode, are planned. The coordination holds well for different faults with no DG connection. For example, in the case of a three-phase (LLL) close in fault of 1551 A at the feeder LF1-1, the recloser fast curve operates in 70 ms, whereas minimum melt time (MMT) of the fuse is 290 ms, as shown in Figure 11.3. The relay current and operating times are shown both as points on the curves and as text within the description boxes.

However, when DG is connected to the test system then, depending upon the DG size, type and location, and the nature and location of the fault, there is a real chance of RF miscoordination, that is, fuses blowing, even in the case of temporary faults. It can be noticed from Figure 11.3 that the chances of RF miscoordination are high when the fault current is high. Generally, maximum fault current is expected in the case of a bolted LLL fault (henceforth referred to as a LLL fault). The same is true in the case of the test system. So to investigate RFC limitations, that is, the conditions where it is not possible to save the fuse, a LLL fault has been simulated at different lengths of LF1-1 with different DG configurations. The total fault currents and the fault currents seen by the recloser and the fuse obtained by simulating a LLL fault at different lengths of the feeder LF1-1 are shown in Figure 11.4. The results are for different values of M and N, where M and N are the numbers of DG units at DG1 and

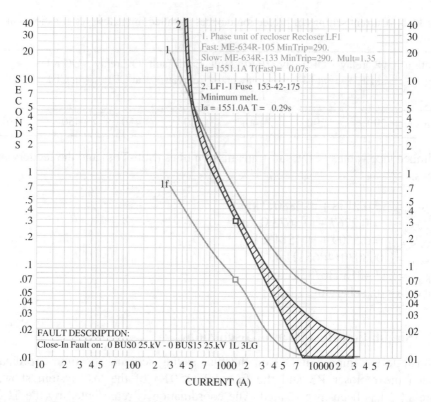

Figure 11.3 Proper coordination between the recloser R3 and the fuse F2 for an LLL (or 3LG with zero ground resistance) close-in fault at LF1-1 without DG. "1f" and "1" stand for the recloser fast and slow characteristics curves, respectively, while "2" represents the fuse characteristics curve

DG2, respectively. It is clear from the figure that the fault currents seen by the recloser R3 and the fuse are different. This is because the recloser is not seeing the fault current contribution from DG2 as the latter is connected downstream of the former whereas this is not true for the fuse, that is, the fuse sees the whole fault current.

The operating times (T_{op}) of the recloser R3 time OC phase element 51, and MMT of the fuse for a LLL fault at different lengths of the feeder LF1-1, without and with DG connection, are shown in Figures 11.5 and 11.6. A coordination time interval (CTI) of 100 ms is assumed to account for CB opening time, errors, and tolerances in current transformers (CTs) and relays. Results for the cases when $M = 1$ or 2 and $N = 10$, are not included as in these cases the conclusions are the same. Moreover, results for single-phase to ground (1LG) or two-phase to ground faults (2LG) are also not considered here as they are cleared by the recloser R3 ground OC time element 51N, without causing RF miscoordination.

A comparison of T_{op} of the recloser R3 with MMTs of the fuse, as shown in Figures 11.5 and 11.6 makes it clear that fuse saving will be ensured by the recloser against all phase faults, that is, phase to phase (LL) and LLL faults, over the entire

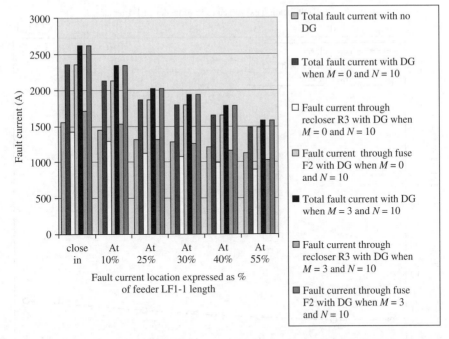

Figure 11.4 The fault current magnitudes for an LLL fault at different lengths of the feeder LF1-1 for different values of M and N

length of the feeder LF1-1 when no DG is connected to the network. When DG is connected to the network, it may not be possible to ensure a CTI of 100 ms between operation of the recloser fast curve and the fuse in the case of LLL faults up to 25% of the feeder LF1-1 length when $M \leq 3$ and $N = 10$. Fuse saving cannot be ensured in that case. From now onwards, LLL faults up to 25% of the feeder LF1-1 length will be mentioned as reference faults and the location corresponding to 25% of the feeder LF1-1 length will be denoted as a reference point (RP).

11.4   Fuse Saving Strategy

As discussed in the previous section that, based on Figures 11.5 and 11.6, a recloser equipped with a 51 element can ensure fuse saving only against low fault currents when DG is present. That is to say, against LL and high impedance LLL faults over the entire length of feeder LF1-1 and against bolted LLL faults only beyond the RP. This is because in the case of low fault currents sufficient time margins are available for a recloser to operate before the fuse has any chance to blow. However, this is not true in the case of high fault currents, for example, bolted faults up to the RP. The increased fault currents together with the downstream location of the DG make it difficult for a recloser to operate before a fuse with a safe time margin.

Modifications of the TCC of the recloser are therefore necessary to save the fuse in the event of temporary faults when DG is connected to the network. One way to

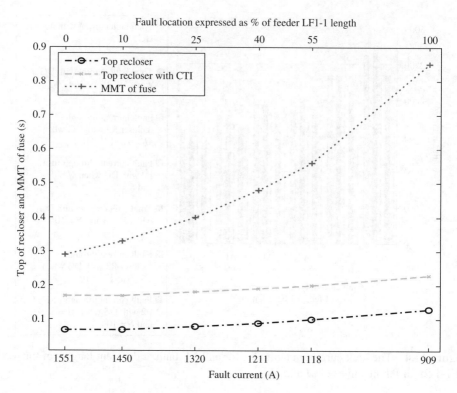

Figure 11.5 T_{op} of the recloser and MMTs of the fuse for a LLL fault at different lengths of the feeder LF1-1 with no DG connection

do this is to follow the procedure described in [12, 13], that is, to modify the fast curve of the recloser on the basis of ratio of I_R and I_F, as mentioned earlier, but, here, this method does not give the desired results. The minimum CTI for the test system is 40 ms in the case of a LLL close in a fault at the feeder LF1-1 when $M = 3$ and $N = 10$. By modifying the R3 recloser fast curve as described in [12, 13], the new minimum CTI for the same fault will be 73 ms, whereas for a fault at 10% of the feeder LF1-1 it will be 96 ms. In both these cases, it is less than the required margin of 100 ms. Moreover, with the application of this method, the whole of the recloser fast curve is modified, which is not desirable in the test system in Figure 11.2 as RF miscoordination is witnessed only over a quarter of the feeder LF1-1, that is, up to the RP. In addition to this, the recloser also sees faults on LF1 where fuse saving is not a matter of concern. In these cases, upgrading the recloser R2 with an instantaneous directional OC element (i.e., 67) having a suitable trip threshold greater than the full load current, that has a total response time (relay operating time plus CB opening time) of 2.5–5 cycles, may not work. As time margins in the scenarios discussed above are very small, the fuse might get damaged before the recloser R2 completes its operation to separate DG2 from the network.

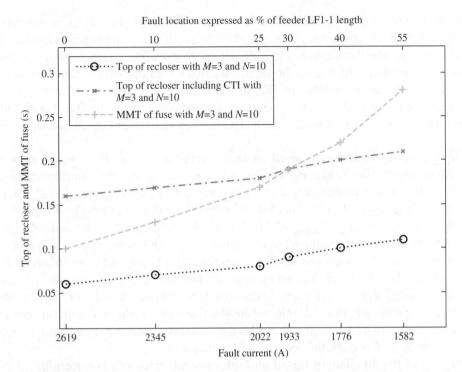

Figure 11.6 T_{op} of the recloser and MMTs of the fuse for faults shown in Figure 11.4 when $M = 3$ and $N = 10$

Alternatively, we propose equipping the recloser with an instantaneous OC phase element (i.e., 50), in addition to the 51 element. This will increase the CTI for the aforementioned faults. In the test system in Figure 11.2, the 50 element in R3, with an operating time of 15 ms when a fault current exceeds its I_{pickup}, will clear all phase faults occurring up to the RP with a reasonable time margin. If the operating time of the 50 element is compared with the fuse blowing times shown in Figures 11.5 and 11.6, it is found that most of the faults will be cleared with a CTI of more than 100 ms. Even in cases where it is not possible to have a CTI of 100 ms, that is, for a LLL close in a fault at LF1-1 when $M = 3$ and $N = 10$, a recloser with a 50 element will ensure a CTI of 85 ms, and for a fault at 10% of the feeder LF1-1 length for the same case,t will ensure a CTI of 115 ms. This shows an improvement in CTI in comparison to results obtained by following the technique described in [12, 13].

11.4.1   Options and Considerations for the Selection of I_{pickup} of the 50 Element

I_{pickup} of the 50 element should be such that it operates only against high magnitude faults that are unlikely to be cleared by a 51 element safely. It is undesirable to set the 50 element to operate against the low magnitude faults, that is, faults occurring beyond the RP. If set too low, it may maltrip when the current momentarily rises in some conditions without a fault, for example, due to transformer inrush current or cold load start conditions. Its setting should be such that it operates only for faults up to the RP, irrespective of the number of DG units connected. $I_{\text{pick-up}}$ of the 50 element can be based either on fault currents calculated without considering the DG contribution or on fault currents determined with the DG contribution, as discussed below:

1. **I_{pickup} of the 50 element based on fault currents with no DG connection:** If the setting of the 50 element is done without considering the contribution from DG, then it will underreach, that is, the area protected by the relay will reduce if, subsequently, DG is connected downstream of the recloser. For example, in the case under study, I_{pickup} of 1290 A for the 50 element will ensure that the recloser will operate for the reference faults without DG connection, as can be seen from Figure 11.5. With this setting, when the DG is connected downstream of the recloser, then the latter, depending upon the number of DG units connected, may not operate for all the reference faults, that is, it will operate only for faults occurring below the RP. For example, when $M = 0$ and $N = 2$, the fault current seen by the recloser for a LLL fault at the RP is only 1262 A, whereas the total fault current is 1486 A. In this case, the recloser will underreach.

2. **I_{pickup} of the 50 element based on fault currents with DG connection:** If the setting of the 50 element is done with the DG contribution taken into account, then it will overreach if, subsequently, DG units which are connected downstream of the recloser are disconnected from the circuit. For example, in the case under study, I_{pickup} of 1095 A for the 50 element will ensure that the relay will operate for reference faults with DG connection. With this setting when downstream-connected DG units are disconnected, the 50 element, depending upon the number of DG units disconnected, may operate for faults occurring beyond the RP and thus encroach onto the area of the 51 element operation. For example, when all the DG units are disconnected, the 50 element with I_{pickup} of 1095 A will operate for faults up to 55% of the feeder LF1-1, as can be seen from Figure 11.5, that is, it will overreach.

Thus, a fixed I_{pickup} of the 50 element based on a fault current with either no DG connection or DG connection, has its limitations. To overcome this problem, an adaptive setting of the 50 element needs to be adopted. The strategy is to set the reference (i.e., initial) I_{pickup} of the element on the basis of a fault current with either no DG connection or with DG connection, and later adaptively change the setting when a DG unit is connected or disconnected from the network.

In this chapter, the initial I_{pickup} of the 50 element is based on the fault current calculated at the RP in the case of a LLL fault with no DG connection. This setting is adaptive, that is, it will change whenever a DG unit is added or removed from the network, to ensure that the recloser with the 50 element operates for all reference faults with a CTI of 100 ms or more. Keeping in mind the proposed adaptive change in settings, a multifunctional microprocessor-based recloser is necessary so that it can be programmed as required. Therefore, replacement of the existing recloser with a microprocessor-based recloser, with communication capabilities, that incorporates both the instantaneous and time OC elements is recommended. A microprocessor-based recloser that is capable of communicating with DG can adaptively change its settings online in accordance with different DG configurations.

11.4.2 Adaptive Algorithm

To enable the 50 element to adaptively modify its I_{pickup}, an algorithm is developed that can be programmed into the recloser control logic together with some essential data that can be stored in a database. The algorithm is shown in Figure 11.7 in the form of a flow chart. In the flow chart, M and N stand for the number of DG units connected at DG1 and DG2, respectively, and I_{set} stands for the reference I_{pickup} of the 50 element. It can have four different values depending upon the values of M while $N = 0$. I_{red} is the reduction in fault current flowing through the recloser R3 in the case of faults at the feeder LF1-1 when a single DG unit is connected at DG2; its value is different for different combinations of M and N. I_p stands for I_{pickup} of the 50 element when $M \geq 0$ and $N \geq 1$, subscripts a, b, c, and d are defined on the basis of I_{red} for various combinations of M and N.

The data that are required to be stored in a database include different possible values of I_{set} and I_{red}. I_{set} is calculated on the basis of the fault currents that result from an LLL fault at 25% of the feeder LF1-1 length (RP) for different combinations of M and N. To implement this algorithm, a communication link is necessary between the recloser and various DG units, so that the former can update its setting when the latter are connected or disconnected from the network. This would require modern reclosers with suitable remote communication capabilities.

11.4.2.1 Description of the Algorithm

1. Through fault analysis of the test system, fault currents due to a LLL fault at different lengths of the feeder LF1-1 with no DG connection, that is, $M = 0$ and $N = 0$, are found and corresponding operating times of the 51 element and fuse are noted. The same step is repeated with all DG units connected and the fault location farthest from the fuse end of the feeder, where it is not possible to have a CTI of 100 ms with a 51 element, is selected as a RP (in the test system in Figure 11.2 this is at 25% of the feeder LF1-1 length). This acts as a basis for selecting I_{pickup} of the 50 element.
2. Similarly, fault currents at the RP are calculated for different values of N and M. For each value of M, I_{set} will be equal to the fault current at the RP when $N = 0$; in the test system, for example, $I_{set} = 1290$ A when $M = 0$, $I_{set} = 1365$ A when $M = 1$,

as shown in Figure 11.7. The 50 element pick-up current is set to be equal to I_{set} when $N = 0$.

3. As N increases, the fault current level changes (reduces on the feeder LF1-1 as seen by the recloser R3 in the tests system in Figure 11.2) and accordingly the pick-up current of the 50 element needs to be adjusted. When the value of N changes, I_{pickup} of the 50 element is reset adaptively by finding the corresponding reduction or increase in utility fault current contribution (I_{red}) flowing through the recloser R3 in the case of an LLL fault occurring at the RP, and then by deducting or adding it from/to the reference setting, (i.e., I_{set}). Thus, when a DG unit is connected or disconnected at DG2, I_{pickup} of the 50 element also decreases or increases accordingly. Figure 11.7 suggests suitable values of I_{red} for the test system that were determined based on fault analysis results.

Figure 11.8 shows the new TCC for the 50 and 51 elements of the recloser R3 of the test system obtained in accordance with the proposed algorithm for the cases when $M = 0$ and $N = 0$ or 10, and when $M = 3$ and $N = 0$ or 5. As can be noticed from these figures I_{pickup} of the 50 element will be different for each of these cases, that is, it changes according to the values of M and N. It will be 1290 A when $M = 0$ and $N = 0$ while it will be 1095 A when $M = 0$ and $N = 10$. Similarly, it will be 1500 A when $M = 3$ and $N = 0$ while it will be 1370 A when $M = 3$ and $N = 5$.

To evaluate the effectiveness of the proposed scheme, a LLL close-in fault and a LLL fault at 10% of feeder LF1-1 length were modeled with the recloser R3 settings modified in accordance with the proposed algorithm when $M = 3$ and $N = 10$. Both of the faults were cleared by the recloser 50 element with a CTI of 85 and 115 ms, respectively. The simulation results for the 10% case are shown in Figure 11.9.

Figure 11.10 shows the fault currents, corresponding T_{op} of the 50 and 51 elements, and MMTs of the fuse in the case of a LLL fault at the RP when $M = 3$ and $N = 10$. For the results shown in the figure, the settings of the 50 element were obtained by implementing the proposed algorithm in MATLAB® while the required input data for the algorithm was found by simulating a LLL fault at the RP with $M = 3$ and $N = 10$. Thus, from the results presented in Figures 11.9 and 11.10, it is clear that the proposed algorithm is valid and the recloser with I_{pickup} modified in accordance with the proposed algorithm can ensure fuse saving with an acceptable time margin in the presence of DG, even in the worst fault conditions.

11.5 How Reclosing Will Be Applied

Ideally, whenever the recloser R3 operates due to a trigger from either the 51 or the 50 elements, a trip signal should be sent simultaneously to reclosers R2 to disconnect DG1 and DG2 from the network to avert formation of an undesirable island. This is in accordance with IEEE standard 1547 that requires DG to be disconnected

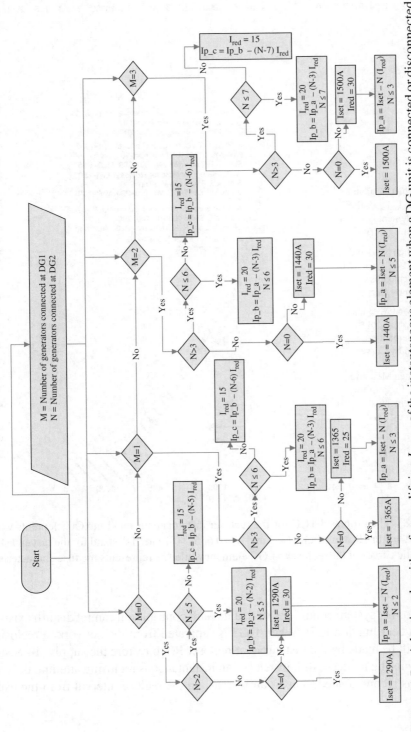

Figure 11.7 Adaptive algorithm for modifying I_{pickup} of the instantaneous element when a DG unit is connected or disconnected from the network

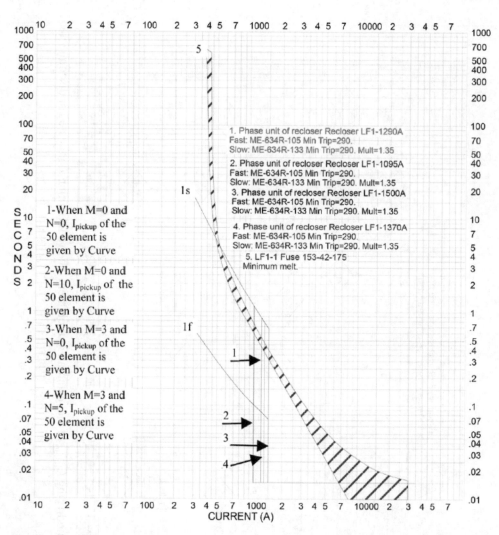

Figure 11.8 The proposed TCC for the recloser instantaneous and time OC elements when $M = 0$ and $N = 0$ or 10, and when $M = 3$ and $N = 0$ or 5. "1f" and "1s" stand for the fast and slow characteristic curves of the recloser 51 OC element while "5" represents the fuse characteristic curve

under abnormal system conditions so that its fault contribution cannot disturb existing protection coordination [25]. When DG2 is separated from the network, a reclosing attempt will be made by the fast curve of recloser R3 to restore the supply. To ensure that the reclosing is successful, instead of an instantaneous reclosing attempt, delayed reclosing is recommended, that is, an increase in the reclose interval from the usual

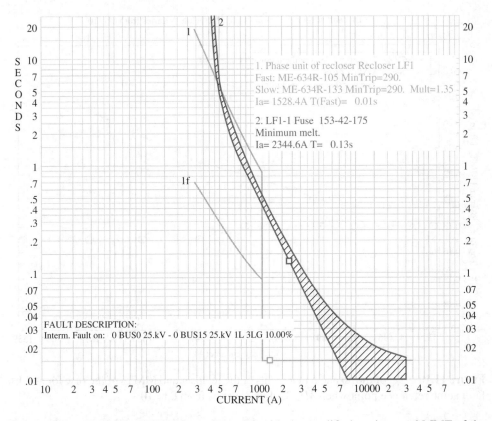

Figure 11.9 Operating time of the recloser R3 with the modified settings and MMT of the fuse F2 for an LLL fault at 10% of the feeder LF1-1 length when $M = 3$ and $N = 10$. "1f" and "1" stand for the recloser fast and slow characteristics curves, respectively, while "2" represents the fuse characteristics curve

300 ms to 1 s should be considered. If the attempt is successful, DG2 will be reconnected to the system with proper synchronization. If the attempt is unsuccessful, then if the fault is at the feeder LF1-1, it will be cleared by the fuse. If it is at the feeder LF1, then it will be cleared by the recloser R3 slow curve. When the recloser R3 closes for the first time and the fault is cleared, its settings will be updated according to the value of M at DG1, with $N = 0$. When, subsequently, the DG2 units are reconnected, the pick-up current of the 50 element will need to be changed according to the proposed algorithm.

11.6 Observations

Although the protection strategy is described for a particular network, it may be successfully applied to restore RFC in other networks. The algorithm for adaptively

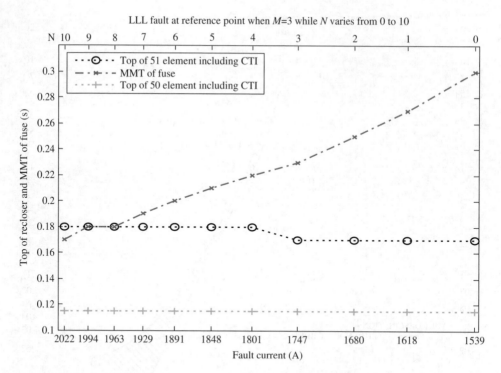

Figure 11.10 T_{op} of the 50 and 51 elements of the recloser and MMTs of the fuse for an LLL fault at the reference point when $M = 3$ while N varies from 0 to 10

changing I_{pickup} of a recloser in response to the varying network configuration can be derived by following the method introduced in Section 11.4.

The proposed algorithm is based on the magnitude of the fault current and is independent of the fault type. The fault current magnitude, irrespective of the nature of the fault, which is most likely to cause the RF miscoordination, is used as the basis for setting the reference I_{pickup} of the 50 element. In the system under study, the maximum fault current that causes RF miscoordination is due to an LLL fault, which is used as a reference in the proposed scheme. As an instantaneous relay can be set to operate in less than a cycle (15 ms in our study), the proposed scheme should be equally effective for both synchronous and induction distributed generators.

The working of the proposed strategy is examined when the contribution from the DG is high compared to the grid source, which will be very rare, as discussed earlier. This is because protection design needs to be done for the worst case scenario. RF miscoordination also occurs even when the DG capacity is less than or equal to the utility supply substation and can be restored by the proposed scheme.

Protection schemes requiring communication links between various protection devices are unpopular due to their high cost. However, by making use of low cost wireless communication channels, which are now commercially available, such schemes may be implemented in a cost effective manner. Moreover, multifunctional microprocessor-based reclosers are already in use for power system protection. So, this scheme could be practical and may be successfully implemented in the actual power networks.

11.7 Conclusions

RFC designed for fuse saving in a DN may be lost when distributed generation is connected to that network, as observed in the case study. To restore this coordination in the case of a typical DN with DG connection, use of a recloser having a 50 element along with a 51 element is shown to be a valid solution as either of them used alone has its limitations. To this end, replacement of the existing recloser with a multifunction microprocessor-based recloser having communication capabilities is recommended. Moreover, it may not be possible to ensure fuse saving with different DG sizes and configurations if the 50 element has a fixed I_{pickup}. To compensate for this, a novel algorithm is proposed for an online adaptive change of I_{pickup} of the 50 element in response to different DG configurations. The simulation results, obtained by adaptively changing relay settings in response to changing DG configurations, confirm that the settings selected theoretically hold well in operation. The proposed scheme combines the characteristics of both the 51 and 50 elements, together with an algorithm for adaptively changing the setting of the 50 element in response to a varying network configuration (due to connection and disconnection of DG units), to ensure fuse saving in the presence of DG.

References

1. Anderson, P.M. (1998) *Power System Protection*, IEEE Press, New York.
2. Dugan, R.C. and Rizy, D.T. (1984) Electric distribution protection problems associated with the interconnection of small, dispersed generation devices. *IEEE Transactions on Power Apparatus and Systems*, **PAS-103** (6), 1121–1127.
3. Girgis, A.A. and Brahma, S.M. (2001) Effect of distributed generation on protective device coordination in distribution system. Proceedings of the Large Engineering Systems Conference on Power Engineering, Halifax, NS, pp. 115–119.
4. Hadjsaid, N., Canard, J.F. and Dumas, F. (1999) Disperse generation impact on distribution network. *IEEE Computer Application in Power*, **12** (2), 22–28.
5. Lu, Y., Hua, L., Wu, J. *et al.*, (2007) A study on the effect of dispersed generator capacity on power system protection. IEEE Power Engineering Society General Meeting, pp. 1-6.
6. Chaitusaney, S. and Yokoyama, A. (2008) Prevention of reliability degradation from recloser-fuse miscoordination due to distributed generation. *IEEE Transaction on Power Delivery*, **23** (4), 2545–2554.
7. Farzanehrafat, A., Javadian, S.A.M., Bathaee, S.M.T., and Haghifam, M.-R. (2008) Maintaining the recloser-fuse coordination in distribution systems in the presence of DG by determining DG's size. The 9th IET International Conference on Developments in Power System Protection, pp. 124–129.

8. Brahma, S.M. and Girgis, A.A. (2001) Impact of distributed generation on fuse and relay coordination: analysis and remedies. Proceedings of the IAESTED International Conference on Power and Energy Systems, pp. 384–389.

9. Zeineldin, H.H. and El-Saadany, E.F. (2010) Fault current limiters to mitigate recloser fuse miscoordination with distributed generation. 10th IET International Conference on Developments in Power System Protection (DPSP). Managing the Change, Manchester, UK, pp. 1–4.

10. Witte, J.F., Mendis, S.R., Bishop, M.T. and Kischefsky, J.A. (1992) Computer-aided recloser applications for distribution systems. *IEEE Computer Applications in Power*, **5** (3), 27–32.

11. Working Group D3 (2004) Impact of distributed resources on distribution relay protection. Line protection subcommittee of the Power System Relay Committee of the IEEE Power Engineering Society, August 2004.

12. Brahma, S.M. and Girgis, A.A. (2002) Microprocessor-based reclosing to coordinate fuse and recloser in a system with high penetration of distributed generation. IEEE Power Engineering Society Winter Meeting, Vol. 1, pp. 453–458.

13. Zamani, A., Sidhu, T., and Yazdani, A. (2010) A strategy for protection coordination in radial distribution networks with distributed generators. IEEE Power and Energy Society General Meeting, pp. 1–8.

14. McCarthy, C., O'Leary, R., and Staszesky, D. (2008) A new fuse-saving philosophy. DistribuTECH, Tampa, FL, January 2008, pp. 1–7.

15. Baran, M. and El-Markabi, I. (2004) Adaptive over current protection for distribution feeders with distributed generators. Proceedings of the IEEE/PES Power Systems Conference and Exposition, Vol. 2, pp. 715–719.

16. Conde, A. and Vazquez, E. (2007) Operation logic proposed for time overcurrent relays. *IEEE Transactions on Power Delivery*, **22**, 2034–2039.

17. Schaefer, N., Degner, T., Shustov, A. *et al.* (2010) Adaptive protection system for distribution networks with distributed energy resources. 10th IET International Conference on Developments in Power System Protection (DPSP), Managing the Change, Manchester, UK, pp. 1–5.

18. Brahma, S.M. and Girgis, A.A. (2004) Development of adaptive protection scheme for distribution systems with high penetration of distributed generation. *IEEE Transactions on Power Delivery*, **19** (1), 56–63.

19. Javadian, S.A.M., Haghifam, M.R., and Barazandeh, P. (2008) An adaptive over-current protection scheme for MV distribution networks including DG. Proceedings of the ISIE08 – IEEE International Symposium on Industrial Electronics, Cambridge, UK, pp. 2520–2525.

20. Viawan, F.A., Karlsson, D., Sannino, A. *et al.* (2006) Protection scheme for meshed distribution systems with high penetration of distributed generation. Power Systems Conference: Advanced Metering, Protection, Control, Communication, and Distributed Resources, pp. 99-104.

21. Tuitemwong, K. and Premrudeepreechacharn, S. (2011) Expert system for protection coordination of distribution system with distributed generators. *International Journal of Electrical Power and Energy Systems*, **33** (3), 466–471.

22. Wan, H., Li, K.K. and Wong, K.P. (2010) An adaptive multiagent approach to protection relay coordination with distributed generators in industrial power distribution system. *IEEE Transactions on Industry Applications*, **46** (5), 2118–2124.
23. Dick, E.P. and Narang, A. (2005) Canadian Urban Benchmark Distribution Systems. Report # CETCVarennes 2005-121 (TR), CANMET Energy Technology Centre, Varennes.
24. IEEE (1987) IEEE Std C37.2-1987. IEEE Standard Electrical Power System Device Function Numbers, IEEE Press, pp. 0–1.
25. IEEE (2003) IEEE Std.1547-2003. IEEE Standard for Interconnecting Distributed Resources with Electric Power Systems, IEEE Press.

Appendix A

SVM for the NPC Converter–MATLAB®-Simulink Models

Sections A.1 and A.2 of this appendix provide supplementary material on NV (nearest-vector) strategies for the NPC (neutral point clamped) converter, while Section A.3 presents representative models/implementations of calculations, converters, and modulation algorithms in MATLAB®-Simulink.

A.1 Calculation of Duty Cycles for Nearest Space Vectors

As explained in Section 5.3, the SV diagram of a three-level converter can be divided into six sextants. Each consists of four small triangles, tr_1–tr_4, illustrated in Figure A.1. For NV strategies, the duty cycles of space vectors are derived from Equation 2.15, where $\mathbf{V}_1, \mathbf{V}_2, \dots, \mathbf{V}_n$, are the NVs for each small triangle. Solution of the above equations does not distinguish between small (also zero) vectors that share the same position on the SV plane. Thus, the derived values of d_{S0}, d_{S1}, and d_Z can be arbitrarily distributed between $\{\mathbf{S0}_1, \mathbf{S0}_2\}$, $\{\mathbf{S1}_1, \mathbf{S1}_2\}$, and $\{\mathbf{Z}_1, \mathbf{Z}_2, \mathbf{Z}_3\}$, respectively.

The expressions for the duty cycles of NVs are taken from [1], and include a correction in Equation A2. It is noted that for the purposes of this section, angle θ stands for the angle of $\mathbf{V}_{REF}$ w.r.t. $\mathbf{L0}$, in the sextant where $\mathbf{V}_{REF}$ belongs (see Figure A.1).

Triangle tr_1: $\{\mathbf{S0}_1, \mathbf{S0}_2\}$, $\mathbf{M}$, $\mathbf{L0}$

$$d_{S0,tr1} = 2 - m(\sqrt{3}\cos\theta + \sin\theta) \tag{A1}$$

$$d_{M,tr1} = 2m\sin\theta \tag{A2}$$

$$d_{L0,tr1} = -1 + m(\sqrt{3}\cos\theta - \sin\theta) \tag{A3}$$

Power Electronic Converters for Microgrids, First Edition. Suleiman M. Sharkh, Mohammad A. Abusara, Georgios I. Orfanoudakis and Babar Hussain.
© 2014 John Wiley & Sons, Ltd. Published 2014 by John Wiley & Sons, Ltd.
Companion Website: www.wiley.com/go/sharkh

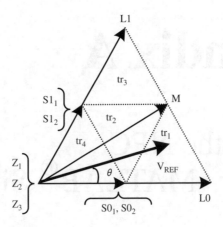

Figure A.1   Space vectors and triangles tr_1–tr_4, in one sextant of a three-level converter

Triangle tr_2: $\{S0_1, S0_2\}$, $\{S1_1, S1_2\}$, M

$$d_{S0,tr2} = 1 - 2m \sin \theta \tag{A4}$$

$$d_{S1,tr2} = 1 + m(\sin \theta - \sqrt{3} \cos \theta) \tag{A5}$$

$$d_{M,tr2} = -1 + m(\sin \theta + \sqrt{3} \cos \theta) \tag{A6}$$

Triangle tr_3: $\{S1_1, S1_2\}$, M, $L1$

$$d_{S1,tr3} = 2 - m(\sqrt{3} \cos \theta' + \sin \theta') \tag{A7}$$

$$d_{M,tr3} = 2m \sin \theta' \tag{A8}$$

$$d_{L1,tr3} = -1 + m(\sqrt{3} \cos \theta' - \sin \theta') \tag{A9}$$

where $\theta' = \pi/3 - \theta$.

Triangle tr_4: $\{Z_1, Z_2, Z_3\}$, $\{S0_1, S0_2\}$, $\{S1_1, S1_2\}$

$$d_{Z,tr3} = 1 - m(\sqrt{3} \cos \theta + \sin \theta) \tag{A10}$$

$$d_{S0,tr3} = m(\sqrt{3} \cos \theta - \sin \theta) \tag{A11}$$

$$d_{S1,tr3} = 2m \sin \theta \tag{A12}$$

A.2   Symmetric Modulation Strategy

This section describes the symmetric modulation strategy, presented in [2] and discussed in Chapter 5. The switching constraints and switching sequences for the first sextant of this strategy are presented in Table A.1.

Table A.1 Duty cycle distribution factors and switching sequences for the symmetric strategy [2]

Triangle	x_{S0}	x_{S1}	Switching sequence	Steps
tr_1	Free	N/A	100-200-210-211-210-200-100	6
tr_{2L}	Free	-1	100-110-210-211-210-110-100	6
tr_{2H}	-1	Free	110-210-211-221-211-210-110	6
tr_3	N/A	Free	110-210-220-221-220-210-110	6
tr_{4L}	Free	-1	100-110-111-211-111-110-100	6
tr_{4H}	-1	Free	110-111-211-221-211-111-110	6

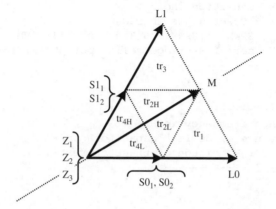

Figure A.2 Space vectors and small triangles for the symmetric strategy

It can be observed that all switching constraints refer to triangles tr_2 and tr_4 in Figure A.2. Namely, in the higher/lower half of these triangles, the distribution factor x_{S0}/x_{S1}, respectively, can only take the value of -1. As explained in Section 5.3.2, this restriction affects the amplitude of NP voltage ripple for $m < 1$. However, it is imposed in order to create a set of switching sequences, each of which comprises six switching steps.

A.3   MATLAB®-Simulink Models

This section includes the following:

1. MATLAB® code for numerical calculation of (normalized) DC-link capacitor rms current and peak–peak voltage ripple for NPC and CHB (cascaded H-bridge) inverters.
2. Simulink model of the "vector selection" block used by SVM (space vector modulation) for the NPC converter.

3. Simulink model of the NPC converter with a NPS (network power switch) circuit.
4. Simulink model of NPC converter modulated with the band-NTV (nearest-three-vector) strategy.
5. MATLAB® code embedded to the "band-NV PWM (pulse width modulation) generator" block in the Simulink model of an NPC converter modulated with the band-NTV strategy (implementing the band-NV algorithm).

Point 1 MATLAB® code for numerical calculation of (normalized) DC-link capacitor rms current and peak–peak voltage ripple for NPC and CHB inverters.

```
%INPUTS
%-------------------------------
M = 2/sqrt(3); % Modulation index
phi = 30/180*pi; % Power angle in rads
strategy = 2; % 1:SPWM, 2:SPWM+1/6 3rd harm., 3: SVM
HBperPh = 2; % Number of H-bridge cells (per phase) for the ML CHB
%-------------------------------


%NORMALISATION
%-------------------------------
f = 1; % Reference frequency in Hz
C = 1; % Capacitance of dc-link capacitors in F
Io_mag = sqrt(2); % Magnitude of phase current in A
%-------------------------------


%CALCULATION
%-------------------------------
dTheta = 2*pi/180;
count = 0;
i_d_NPC_rmssq = 0; % i.e. i_d_NPC_rms square
i_d_NPC_dc = 0;
i_d_CHB_rmssq = 0; % i.e. i_d_CHB_rms square
i_d_CHB_dc = 0;
i_d_NPC = zeros(1);
i_d_CHB = zeros(1);
i_d_CHB_rmssq_ml = zeros(10);
i_d_CHB_dc_ml = zeros(10);
i_d_CHB_ml = zeros(1,1);
duty_ml = zeros(1);

for theta = 0:dTheta:2*pi-dTheta
    count = count + 1;

    %Duty cycle calculation
    dutyA = M*cos(theta);
    dutyB = M*cos(theta - 2*pi/3);
    dutyC = M*cos(theta + 2*pi/3);

    cmSignal = 0; % Common-mode voltage reference signal
```

```
        if strategy == 2
            cmSignal = -1/6*M*cos(3*theta);
        else if strategy == 3
                dutyABC = [dutyA dutyB dutyC];
                cmSignal = -0.5*(max(dutyABC) + min(dutyABC));
            end
        end
        dutyA = dutyA + cmSignal;
        dutyB = dutyB + cmSignal;
        dutyC = dutyC + cmSignal;

        %Three-phase currents
        iA = Io_mag*cos(theta-phi);
        iB = Io_mag*cos(theta-phi-2*pi/3);
        iC = Io_mag*cos(theta-phi+2*pi/3);
        I3ph = [iA iB iC];

        %NPC inverter calculations
        [dutySorted, indices] = sort([dutyA*(dutyA > 0) dutyB*(dutyB > 0)
    dutyC*(dutyC > 0)], 'ascend');

        duty1 = dutySorted(1);
        duty2 = dutySorted(2);
        duty3 = dutySorted(3);
        i1 = I3ph(indices(1));
        i2 = I3ph(indices(2));
        i3 = I3ph(indices(3));
        i_d_NPC_rmssq = i_d_NPC_rmssq + (duty2-duty1)*(i2+i3)^2 +
    (duty3-duty2)*(i3)^2;
        i_d_NPC(count) = iA*dutyA*(dutyA > 0) + iB*dutyB*(dutyB > 0) +
    iC*dutyC*(dutyC > 0);
        i_d_NPC_dc = i_d_NPC_dc + i_d_NPC(count);

        %CHB inverter calculations
        i_d_CHB_rmssq = i_d_CHB_rmssq + iA^2*abs(dutyA);
        i_d_CHB(count) = iA*dutyA;
        i_d_CHB_dc = i_d_CHB_dc + i_d_CHB(count);

        %Multilevel CHB (CHB ML) inverter calculations
        HBhigh = floor(abs(dutyA)*HBperPh); %Num-
    ber of cells with duty cycle = +/-1 (for phase A)
        HBduty = HBperPh * mod(abs(dutyA), 1/HBperPh);

        for n = 1:HBperPh
            if n <= HBhigh
                duty_ml(n) = sign(dutyA); % Duty cycle of the HBhigh
    lower cells
            else
                duty_ml(n) = 0; % Duty cycle of the (HBhigh+2) up to the
    highest-level cell
            end
```

```
        end
        duty_ml(HBhigh + 1) = sign(dutyA)*HBduty; % Duty cycle of the
(HBhigh+1)-level cell

        for n = 1:HBperPh
            i_d_CHB_rmssq_ml(n) = i_d_CHB_rmssq_ml(n) +
iA^2*abs(duty_ml(n));
            i_d_CHB_ml(count,n) = iA*duty_ml(n);
            i_d_CHB_dc_ml(n) = i_d_CHB_dc_ml(n) + i_d_CHB_ml(count,n);
        end

end

%RESULTS
%-------------------------------

%NPC inverter results
fprintf('\n\nNPC\n-----------------');
i_d_NPC_rmssq = i_d_NPC_rmssq*dTheta/(2*pi);
i_d_NPC_dc = i_d_NPC_dc*dTheta/(2*pi);
i_C_NPC_rms = sqrt(i_d_NPC_rmssq - i_d_NPC_dc^2) %Capacitor rms cur-
rent for NPC

i_C_NPC = i_d_NPC - i_d_NPC_dc;
dQ_C_NPC = i_C_NPC.*dTheta/(2*pi*f);
Q_C_NPC = cumsum(dQ_C_NPC);
Q_C_NPCpp = max(Q_C_NPC) - min(Q_C_NPC);
V_C_NPCpp = Q_C_NPCpp/C %Capacitor peak-peak voltage ripple for NPC

%CHB inverter results
fprintf('\n\nCHB\n-----------------');
i_d_CHB_rmssq = i_d_CHB_rmssq*dTheta/(2*pi);
i_d_CHB_dc = i_d_CHB_dc*dTheta/(2*pi);
i_C_CHB_rms = sqrt(i_d_CHB_rmssq - i_d_CHB_dc^2) %Capacitor rms cur-
rent for CHB

i_C_CHB = i_d_CHB - i_d_CHB_dc;
dQ_C_CHB = i_C_CHB.*dTheta/(2*pi*f);
Q_C_CHB = cumsum(dQ_C_CHB);
Q_C_CHBpp = max(Q_C_CHB) - min(Q_C_CHB);
V_C_CHBpp = Q_C_CHBpp/C %Capacitor peak-peak voltage ripple for CHB

%CHB ML inverter results
for n = 1:HBperPh
    fprintf('\n\nML CHB, Cell %d \n-----------------',n);
    i_d_CHB_rmssq_ml(n) = i_d_CHB_rmssq_ml(n)*Dtheta/(2*pi);
    i_d_CHB_dc_ml(n) = i_d_CHB_dc_ml(n)*Dtheta/(2*pi);
    i_C_CHB_rms_ml = sqrt(i_d_CHB_rmssq_ml(n) - i_d_CHB_dc_ml(n)^2)

    i_C_CHB_ml = i_d_CHB_ml(:,n) - i_d_CHB_dc_ml(n);
    dQ_C_CHB_ml = i_C_CHB_ml.*Dtheta/(2*pi*f);
```

```
      Q_C_CHB_ml = cumsum(dQ_C_CHB_ml);
      Q_C_CHBpp_ml = max(Q_C_CHB_ml) - min(Q_C_CHB_ml);
      V_C_CHBpp_ml = Q_C_CHBpp_ml/C
end
```

Point 2 Simulink model of the "vector selection" block used by SVM for the NPC converter is shown in Figure A.3.

Point 3 Simulink model of NPC converter with NPS circuit is shown in Figure A.4.

Point 4 Simulink model of NPC converter modulated with the band-NTV strategy is shown in Figure A.5.

Point 5 MATLAB® code embedded to the "band-NV PWM generator" block in the Simulink model of an NPC converter modulated with the band-NTV strategy.

```
% NOTE: The code below operates through an inter-
face to the "Band-NV PWM generator" block.

function [vNum,i_M,i_Spos,i_Sneg,tprev_new,Vbeg,ContInt,V_C1_ref,sM,
UI_Flag,UIsign,t_sM,bal_Flag,balOK,Region,sV_C1,V_Bandpk,
DQcross_flag,Vector1,Vector2,Vector3,Vector4,Time1,Time2,Time3,Time4,
i_Smax,iNP_BandNTV,x_S1,x_S0] = ...
    fcn(t,fs,tprev,V_C1,Iabc,M,vNum_prev,i_M_prev,i_Spos_prev,
i_Sneg_prev,Vbeg_prev,ContInt_prev,sM_prev,UI_Flag_prev,UIsign_prev,
t_sM_prev,bal_Flag_prev,balOK_prev,Region_prev,sV_C1_prev,
V_Bandpk_prev,DQcross_flag_prev,V_C1_ref_prev,Vector1_prev,
Vector2_prev,Vector3_prev,Vector4_prev,Time1_prev,Time2_prev,
Time3_prev,Time4_prev,i_Smax_prev,iNP_BandNTV_prev,x_S1_prev,
x_S0_prev)

%INPUT
f = 50;
Vnom = 900;
C = 0.5e-3;
% x_max = 1;
notUsing8stepsForSmallBalancing = 1;

%Hold the values below, unless they change later
Vector1 = Vector1_prev;
Vector2 = Vector2_prev;
Vector3 = Vector3_prev;
Vector4 = Vector4_prev;

Time1 = Time1_prev;
Time2 = Time2_prev;
Time3 = Time3_prev;
Time4 = Time4_prev;

vNum = vNum_prev;
i_M = i_M_prev;
```

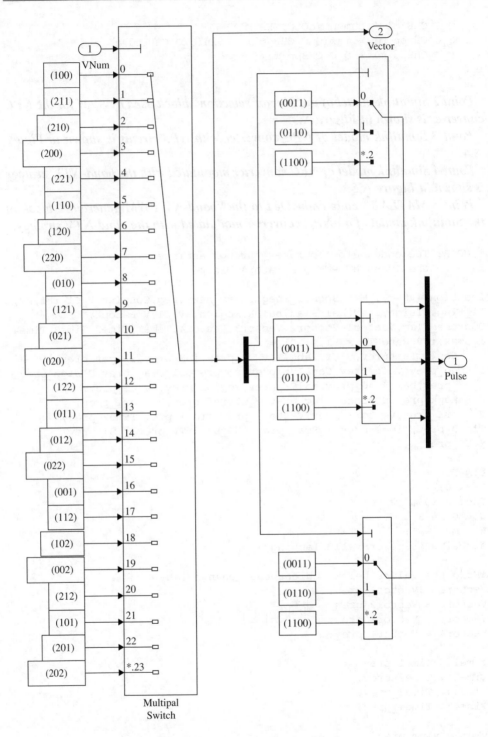

Figure A.3   Vector selection block

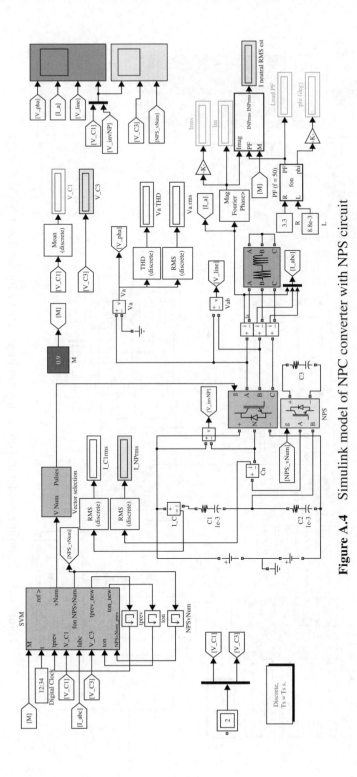

Figure A.4   Simulink model of NPC converter with NPS circuit

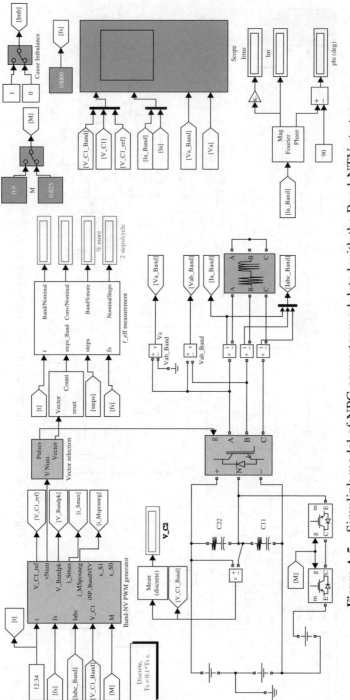

Figure A.5   Simulink model of NPC converter modulated with the Band-NTV strategy

```
i_Spos = i_Spos_prev;
i_Sneg = i_Sneg_prev;
i_Smax = i_Smax_prev;
iNP_BandNTV = iNP_BandNTV_prev;
x_S0 = x_S0_prev;
x_S1 = x_S1_prev;

UI_Flag = UI_Flag_prev;
UIsign = UIsign_prev;
t_sM = t_sM_prev;
DQcross_flag = DQcross_flag_prev;
bal_Flag = bal_Flag_prev;
balOK = balOK_prev;
Region = Region_prev;
V_Bandpk = V_Bandpk_prev;
Vbeg = Vbeg_prev;
V_C1_ref = V_C1_ref_prev;
ContInt = ContInt_prev;
sV_C1 = sV_C1_prev;
sM = sM_prev;

Tsw = 1/fs;
w = 2*pi*f;
ref = cos(w*t);

% if M < 0.95
%      mult = 1.07;
% else
%      mult = 1.01;
% end
mult = 1;

if t > tprev+Tsw
    %New switching period!
    tprev_new = t;

    % theta is calculated based on tprev,
    % therefore remains the same during a switching period
    theta = mod(w*tprev, 2*pi);
    sextantNum = floor(theta / (pi/3));
    theta = mod(theta, pi/3); % Working with the first Sextant only!
    theta_tr = acos(1/(2*M))-pi/6;

    %Duty cycles
    dS0mod = dS0(theta, M, 1);
    dS1mod = dS1(theta, M, 1);
    dMmod = dM(theta, M, 1);

    %Vector selection
    vNumS0 = mod(4*sextantNum, 24); %S0,1->+0, S0,2->+1
```

```
    vNumS1 = mod(4*sextantNum + 4, 24); %S1,1->+4,
S1,2->+5
    vNumM = mod(4*sextantNum + 2, 24);

    %Calculating i_S0, i_S1
    IofS0 = 0;
    IofS1 = 0;

    if vNumS0 == 0 || vNumS0 == 12
        IofS0 = Iabc(1);
    else if vNumS0 == 4 || vNumS0 == 16
            IofS0 = Iabc(3);
            else if vNumS0 == 8 || vNumS0 == 20
            IofS0 = Iabc(2);
                end
        end
    end

    if vNumS1 == 0 || vNumS1 == 12
        IofS1 = Iabc(1);
    else if vNumS1 == 4 || vNumS1 == 16
            IofS1 = Iabc(3);
            else if vNumS1 == 8 || vNumS1 == 20
            IofS1 = Iabc(2);
                end
        end
    end

    i_S0 = dS0mod*IofS0;
    i_S1 = dS1mod*IofS1;
    i_Smax = abs(i_S0) + abs(i_S1);

    % Deriving i_Spos|Str and i_Sneg|Str
    i_Spos = i_Smax;
    i_Sneg = -i_Smax;

    %Calculating i_M
    IofM = 0;

    if vNumM == 2 || vNumM == 14
        IofM = Iabc(2);
    else if vNumM == 6 || vNumM == 18
            IofM = Iabc(1);
            else if vNumM == 10 || vNumM == 22
                IofM = Iabc(3);
                end
        end
    end

    i_M = dMmod*IofM;
```

```
% CALCULATION OF V_C1_ref
%--------------------------------------------------------------
--------------------
sM = sign(i_M);

%Checking inverter operation Region (0, 1, or 2) and Bal-
ance whenever i_M changes sign
if sM_prev ~= sM && t > t_sM_prev + 5*Tsw

    %Check that balance was not lost
    balOK = 1;
    if bal_Flag == 0
        balOK = 0;
    end

    % Check if there has been an UI between the previ-
ous change of sM and this one
    Region = 1;
    if UI_Flag_prev == 0 % If not, we are in Region 0
        Region = 0;
    end

    %Reinitialize flags for next i_M half cycle (after
some delay, because between quick sign changes
    %there are no UIs and this confuses the code,
sending it to Region 0)
    bal_Flag = 0;
    UI_Flag = 0;
    DQcross_flag = 0;
    t_sM = t;
end

sV_C1 = sign(V_C1 - Vnom);
% Check if V_C1 crossed Vnom between the
previous change of sM and this one
if sV_C1_prev ~= sV_C1
    bal_Flag = 1;
end

% Check if there has been an UI between the
previous change of sM and this one
if i_M > -i_Sneg || i_M < -i_Spos %aDQS < ai_M %Inside an
uncontrollable interval
    UI_Flag = 1; % UI appeared between two sign changes of i_M,
so Region = 1 or 2.
    if i_M > 0
        UIsign = 1; % Sign of i_M during last UI
    else
```

```
            UIsign = -1;
        end
    end

    if balOK_prev == 0 || Region_prev == 0
        ContInt = 1;

        Vbeg = Vnom;
        V_Bandpk = 0;
        V_C1_ref = Vnom;
    else % Region 1, Balanced operation

        % Sample V_C1 at the beginning and at the end of an
uncontrollable interval.
        %Sampling voltage at the beginning and the end of
uncontrollable intervals (if any)
        if ~(i_M > -i_Sneg || i_M < -i_Spos) %Inside a control-
lable interval
            ContInt = 1;

            %At the beginning of a controllable interval:
            %If the previous UI has driven V_C1 to its sign,
sample V_C1.
            %This is the final voltage, after the
uncontrollable int.
            if ContInt_prev == 0 && sV_C1 == UIsign
                %Change Vref
                V_Bandpk = mult * 0.5 * abs(V_C1 - Vbeg_prev);

                %Set the reference (target) voltage
according to the sign (or 0) of i_M
                %in the previous uncontrollable interval.
                V_C1_ref = Vnom + UIsign * V_Bandpk;
            end
        else
            ContInt = 0;

            %At the beggining of an uncontrollable interval
sample V_C1. This is the initial voltage, at the
beggining of the uncontrollable int.
            if ContInt_prev == 1 && DQcross_flag_prev == 0
                Vbeg = V_C1;
                DQcross_flag = 1; % To sample the very
beginning of the uncontrollable interval
            end
        end
    end
```

```
%ADJUSTMENT OF DUTY CYCLES
%------------------------------------------------------------------
----------------------

    %Voltage deviation due to M, S0,1 and S1,1
    DVofM = i_M*Tsw/(2*C);
    DVofS0 = i_S0*Tsw/(2*C);
    DVofS1 = i_S1*Tsw/(2*C);

    vNumS0_1 = vNumS0;
    vNumS0_2 = vNumS0 + 1;

    vNumS1_1 = vNumS1;
    vNumS1_2 = vNumS1 + 1;

    x_S0 = 0;
    x_S1 = 0;

    if theta < theta_tr

        Triangle = 1;

        %Low outer triangle, use S0, M, L0
        vNumL0 = mod(4*sextantNum + 3, 24);
        dL0mod = 1-dS0mod-dMmod;

        %Distribution of duty cycle between S0,1 and S0,2
        %For the NTV, it is actually selection between S0,1 or S0,2
        if DVofS0 == 0
            % Arbitrarily select x_S0 = 1, it doesn't affect vNP
            Vector1 = vNumS0_1; Time1 = dS0mod*Tsw;
            Vector2 = vNumL0; Time2 = dL0mod*Tsw;
            Vector3 = vNumM; Time3 = dMmod*Tsw;
        else
            x_S0 = (V_C1_ref - V_C1 - DVofM)/DVofS0;

            if x_S0 >= 0
                x_S0 = 1;
                Vector1 = vNumS0_1; Time1 = dS0mod*Tsw;
                Vector2 = vNumL0; Time2 = dL0mod*Tsw;
                Vector3 = vNumM; Time3 = dMmod*Tsw;
            else
                x_S0 = -1;
                Vector1 = vNumL0; Time1 = dL0mod*Tsw;
                Vector2 = vNumM; Time2 = dMmod*Tsw;
                Vector3 = vNumS0_2; Time3 = dS0mod*Tsw;
            end
        end

    else if theta < pi/3 - theta_tr
            %Triangle 2
```

```
            Triangle = 2;
            balancing_flag = 0;

            if V_C1_ref > V_C1 + DVofM + abs(DVofS0) + abs(DVofS1)
                %Select the small vectors that cause positive DVs
                if DVofS0 > 0
                    x_S0 = 1;
                else
                    x_S0 = -1;
                end

                if DVofS1 > 0
                    x_S1 = 1;
                else
                    x_S1 = -1;
                end
            else if V_C1_ref < V_C1 + DVofM - abs(DVofS0) -
abs(DVofS1)
                %Select the small vectors that cause
negative DVs
                if DVofS0 > 0
                    x_S0 = -1;
                else
                    x_S0 = 1;
                end

                if DVofS1 > 0
                    x_S1 = -1;
                else
                    x_S1 = 1;
                end
            else
                balancing_flag = 1;
                finalV_C1_a = V_C1 + DVofM + abs(DVofS0) -
abs(DVofS1);
                finalV_C1_b = V_C1 + DVofM - abs(DVofS0) +
abs(DVofS1);

                if abs(V_C1_ref - finalV_C1_a) < abs(V_C1_ref -
finalV_C1_b)
                    %Select S0 that causes positive
DV and S1 that
                    %causes negative DV (a).
                    if DVofS0 > 0
                        x_S0 = 1;
                    else
                        x_S0 = -1;
                    end

                    if DVofS1 > 0
                        x_S1 = -1;
```

```
                        else
                            x_S1 = 1;
                        end

                else
                        %Select S0 that causes
negative DV and S1 that
                        %causes positive DV (b).
                        if DVofS0 > 0
                            x_S0 = -1;
                        else
                            x_S0 = 1;
                        end

                        if DVofS1 > 0
                            x_S1 = 1;
                        else
                            x_S1 = -1;
                        end
                    end
                end
            end

            if notUsing8stepsForSmallBalancing == 1 && balanc-
ing_flag == 1 && x_S0 == 1 && x_S1 == 1
                    x_S0 = -1;
                    x_S1 = -1;
            end

            if x_S0 == 1 && x_S1 == 1 % This is the 8-step sequence
                Vector1 = vNumS0_1; Time1 = dS0mod*Tsw;
                Vector2 = vNumM; Time2 = dMmod*Tsw;
                Vector3 = vNumS1_1; Time3 = dS1mod*Tsw;
            else if x_S0 == 1 && x_S1 == -1
                    Vector1 = vNumS0_1; Time1 = dS0mod*Tsw;
                    Vector2 = vNumS1_2; Time2 = dS1mod*Tsw;
                    Vector3 = vNumM; Time3 = dMmod*Tsw;
                else if x_S0 == -1 && x_S1 == 1
                        Vector1 = vNumM; Time1 = dMmod*Tsw;
                        Vector2 = vNumS0_2; Time2 = dS0mod*Tsw;
                        Vector3 = vNumS1_1; Time3 = dS1mod*Tsw;
                    else
                        Vector1 = vNumS1_2; Time1 = dS1mod*Tsw;
                        Vector2 = vNumM; Time2 = dMmod*Tsw;
                        Vector3 = vNumS0_2; Time3 = dS0mod*Tsw;
                    end
                end
            end
        else
            Triangle = 3;
```

```
            vNumL1 = mod(4*sextantNum + 7, 24);
            dL1mod = 1-dS1mod-dMmod;

            %Distribution of duty cycle between S0,1 and S0,2
            if DVofS1 == 0
                % Arbitrarily select x_S1 = 1, it doesn't affect vNP
                Vector1 = vNumM; Time1 = dMmod*Tsw;
                Vector2 = vNumL1; Time2 = dL1mod*Tsw;
                Vector3 = vNumS1_1; Time3 = dS1mod*Tsw;
            else
                x_S1 = (V_C1_ref - V_C1 - DVofM)/DVofS1;

                if x_S1 >= 0
                    x_S1 = 1;
%                     Vector1 = vNumM; Time1 = dMmod*Tsw;
%                     Vector2 = vNumL1; Time2 = dL1mod*Tsw;
%                     Vector3 = vNumS1_1; Time3 = dS1mod*Tsw;
                    Vector1 = vNumS1_1; Time1 = dS1mod*Tsw;
                    Vector2 = vNumL1; Time2 = dL1mod*Tsw;
                    Vector3 = vNumM; Time3 = dMmod*Tsw;
                else
                    x_S1 = -1;
%                     Vector1 = vNumS1_2; Time1 = dS1mod*Tsw;
%                     Vector2 = vNumM; Time2 = dMmod*Tsw;
%                     Vector3 = vNumL1; Time3 = dL1mod*Tsw;
                    Vector1 = vNumL1; Time1 = dL1mod*Tsw;
                    Vector2 = vNumM; Time2 = dMmod*Tsw;
                    Vector3 = vNumS1_2; Time3 = dS1mod*Tsw;
                end
            end
        end
    end

    iNP_BandNTV = i_M + x_S0*i_S0 + x_S1*i_S1;
else
    %Same switching period
    tprev_new = tprev;
end


%SWITCHING THE CONVERTER
%----------------------------------------------------------------
--------------

if t < tprev_new + (Time1)/2
    vNum = Vector1;
else if t < tprev_new + (Time1 + Time2)/2
        vNum = Vector2;
    else if t < tprev_new + (Time1 + Time2 + 2*Time3)/2
            vNum = Vector3;
        else if t < tprev_new + (Time1 + 2*Time2 + 2*Time3)/2
```

```
                vNum = Vector2;
            else
                vNum = Vector1;
            end
        end
    end
end
```

References

1. Celanovic, N. and Boroyevich, D. (2000) A comprehensive study of neutral-point volt-age balancing problem in three-level neutral-point-clamped voltage source PWM inverters. *IEEE Transactions on Power Electronics*, **15** (2), 242–249.
2. Pou, J., Pindado, R., Boroyevich, D. and Rodriguez, P. (2005) Evaluation of the low-frequency neutral-point voltage oscillations in the three-level inverter. *IEEE Transactions on Industrial Electronics*, **52** (6), 1582–1588.

Appendix B

DC-Link Capacitor Current Numerical Calculation

```
% MATLAB code for numerical calculation of (normalized)
dc-link capacitor
% rms current and peak-peak voltage ripple
for NPC and CHB 3L inverters.

%INPUTS
%-------------------------------
M = 2/sqrt(3); % Modulation index
phi = 30/180*pi; % Power angle in rads
strategy = 2; % 1:SPWM, 2:SPWM+1/6 3rd harm., 3: SVM
HBperPh = 2; % Number of H-bridge cells (per phase) for the ML CHB
%-------------------------------


%NORMALISATION
%-------------------------------
f = 1; % Reference frequency in Hz
C = 1; % Capacitance of dc-link capacitors in F
Io_mag = sqrt(2); % Magnitude of phase current in A
%-------------------------------


%CALCULATION
%-------------------------------
dTheta = 2*pi/180;
count = 0;
i_d_NPC_rmssq = 0; % i.e. i_d_NPC_rms square
i_d_NPC_dc = 0;
i_d_CHB_rmssq = 0; % i.e. i_d_CHB_rms square
i_d_CHB_dc = 0;
i_d_NPC = zeros(1);
```

Power Electronic Converters for Microgrids, First Edition. Suleiman M. Sharkh, Mohammad A. Abusara,
Georgios I. Orfanoudakis and Babar Hussain.
© 2014 John Wiley & Sons, Ltd. Published 2014 by John Wiley & Sons, Ltd.
Companion Website: www.wiley.com/go/sharkh

```
i_d_CHB = zeros(1);
i_d_CHB_rmssq_ml = zeros(10);
i_d_CHB_dc_ml = zeros(10);
i_d_CHB_ml = zeros(1,1);
duty_ml = zeros(1);

for theta = 0:dTheta:2*pi-dTheta
    count = count + 1;

    %Duty cycle calculation
    dutyA = M*cos(theta);
    dutyB = M*cos(theta - 2*pi/3);
    dutyC = M*cos(theta + 2*pi/3);

    cmSignal = 0; % Common-mode voltage reference signal
    if strategy == 2
        cmSignal = -1/6*M*cos(3*theta);
    else if strategy == 3
            dutyABC = [dutyA dutyB dutyC];
            cmSignal = -0.5*(max(dutyABC) + min(dutyABC));
        end
    end
    dutyA = dutyA + cmSignal;
    dutyB = dutyB + cmSignal;
    dutyC = dutyC + cmSignal;

    %Three-phase currents
    iA = Io_mag*cos(theta-phi);
    iB = Io_mag*cos(theta-phi-2*pi/3);
    iC = Io_mag*cos(theta-phi+2*pi/3);
    I3ph = [iA iB iC];

    %NPC inverter calculations
[dutySorted, indices] = sort([dutyA*(dutyA > 0) dutyB*(dutyB > 0)
dutyC*(dutyC > 0)], 'ascend');

    duty1 = dutySorted(1);
    duty2 = dutySorted(2);
    duty3 = dutySorted(3);
    i1 = I3ph(indices(1));
    i2 = I3ph(indices(2));
    i3 = I3ph(indices(3));

    i_d_NPC_rmssq = i_d_NPC_rmssq + (duty2-duty1)*(i2+i3)^2 +
(duty3-duty2)*(i3)^2;
    i_d_NPC(count) = iA*dutyA*(dutyA > 0) + iB*dutyB*(dutyB > 0) +
iC*dutyC*(dutyC > 0);
    i_d_NPC_dc = i_d_NPC_dc + i_d_NPC(count);

    %CHB inverter calculations
    i_d_CHB_rmssq = i_d_CHB_rmssq + iA^2*abs(dutyA);
```

```
    i_d_CHB(count) = iA*dutyA;
    i_d_CHB_dc = i_d_CHB_dc + i_d_CHB(count);

    %Multilevel CHB (CHB ML) inverter calculations
    HBhigh = floor(abs(dutyA)*HBperPh); % Number
of cells with duty cycle = +/-1 (for phase A)
    HBduty = HBperPh * mod(abs(dutyA), 1/HBperPh);

    for n = 1:HBperPh
        if n <= HBhigh
            duty_ml(n) = sign(dutyA); % Duty cycle of the
HBhigh lower cells
        else
            duty_ml(n) = 0; % Duty cycle of the (HBhigh+2) up
to the highest-level cell
        end
    end
    duty_ml(HBhigh + 1) = sign(dutyA)*HBduty; % Duty cycle of
the (HBhigh+1)-level cell

    for n = 1:HBperPh
        i_d_CHB_rmssq_ml(n) = i_d_CHB_rmssq_ml(n) +
iA^2*abs(duty_ml(n));
        i_d_CHB_ml(count,n) = iA*duty_ml(n);
        i_d_CHB_dc_ml(n) = i_d_CHB_dc_ml(n) + i_d_CHB_ml(count,n);
    end

end

%RESULTS
%-------------------------------

%NPC inverter results
fprintf('\n\nNPC\n-----------------');
i_d_NPC_rmssq = i_d_NPC_rmssq*dTheta/(2*pi);
i_d_NPC_dc = i_d_NPC_dc*dTheta/(2*pi);
i_C_NPC_rms = sqrt(i_d_NPC_rmssq - i_d_NPC_dc^2) %Capacitor rms
current for NPC

i_C_NPC = i_d_NPC - i_d_NPC_dc;
dQ_C_NPC = i_C_NPC.*dTheta/(2*pi*f);
Q_C_NPC = cumsum(dQ_C_NPC);
Q_C_NPCpp = max(Q_C_NPC) - min(Q_C_NPC);
V_C_NPCpp = Q_C_NPCpp/C %Capacitor peak-peak voltage ripple for NPC

%CHB inverter results
fprintf('\n\nCHB\n-----------------');
i_d_CHB_rmssq = i_d_CHB_rmssq*dTheta/(2*pi);
i_d_CHB_dc = i_d_CHB_dc*dTheta/(2*pi);
i_C_CHB_rms = sqrt(i_d_CHB_rmssq - i_d_CHB_dc^2) %Capacitor rms
current for CHB
```

```
i_C_CHB = i_d_CHB - i_d_CHB_dc;
dQ_C_CHB = i_C_CHB.*dTheta/(2*pi*f);
Q_C_CHB = cumsum(dQ_C_CHB);
Q_C_CHBpp = max(Q_C_CHB) - min(Q_C_CHB);
V_C_CHBpp = Q_C_CHBpp/C %Capacitor peak-peak voltage ripple for CHB

%CHB ML inverter results
for n = 1:HBperPh
    fprintf('\n\nML CHB, Cell %d \n-----------------',n);
    i_d_CHB_rmssq_ml(n) = i_d_CHB_rmssq_ml(n)*Dtheta/(2*pi);
    i_d_CHB_dc_ml(n) = i_d_CHB_dc_ml(n)*Dtheta/(2*pi);
    i_C_CHB_rms_ml = sqrt(i_d_CHB_rmssq_ml(n) - i_d_CHB_dc_ml(n)^2)

    i_C_CHB_ml = i_d_CHB_ml(:,n) - i_d_CHB_dc_ml(n);
    dQ_C_CHB_ml = i_C_CHB_ml.*Dtheta/(2*pi*f);
    Q_C_CHB_ml = cumsum(dQ_C_CHB_ml);
    Q_C_CHBpp_ml = max(Q_C_CHB_ml) - min(Q_C_CHB_ml);
    V_C_CHBpp_ml = Q_C_CHBpp_ml/C
end
```

Index

Power Electronic Converters for Microgrids, First Edition. Suleiman M. Sharkh, Mohammad A. Abusara, Georgios I. Orfanoudakis and Babar Hussain.
© 2014 John Wiley & Sons, Ltd. Published 2014 by John Wiley & Sons, Ltd.
Companion Website: www.wiley.com/go/sharkh